D0721793

UNBOUNDED LINEAR OPERATORS

UNBOUNDED LINEAR OPERATORS

THEORY AND APPLICATIONS

Seymour Goldberg

Professor of Mathematics
University of Maryland

Dover Publications, Inc.
New York

To

LILLIAN, FLORENCE, and BENJIE

Published in Canada by General Publishing Company, Ltd., 30 Lesmill Road, Don Mills, Toronto, Ontario.
Published in the United Kingdom by Constable and Company, Ltd., 10 Orange Street, London WC2H 7EG.

This Dover edition, first published in 1985, is an unabridged, slightly corrected edition of the work first published by the McGraw-Hill Book Company, Inc., New York, in 1966.

Manufactured in the United States of America
Dover Publications, Inc., 31 East 2nd Street, Mineola, N.Y. 11501

Library of Congress Cataloging in Publication Data

Goldberg, Seymour, 1928–
Unbounded linear operators.

Bibliography: p.
Includes index.
1. Linear operators. I. Title.
QA329.2.G65 1985 515.7′246 84-25952
ISBN 0-486-64830-3 (pbk.)

PREFACE

In recent years important progress has been made in the study of linear operators by extending to closed operators many fundamental theorems which were known for bounded operators. The applications of the theory permit unification of a series of diverse questions in differential equations, which leads to significant results with substantial simplification.

The aim of this book is to present a systematic treatment of the theory of unbounded linear operators in normed linear (not necessarily Hilbert) spaces with applications to differential equations. Except for the last chapter, the material is quite self-contained. The reader is assumed to be familiar with metric spaces and real variable theory.

The book could be introduced in a course in functional analysis, particularly when linear operators are studied. By considering differential or integro-differential equations from the point of view of operator theory, the material may also be useful to those concerned with the more applied aspects of analysis.

Chapter I gives the elementary theory of normed linear spaces and, in particular, Hilbert space, which is used throughout the book. In Chapters II and IV, the basic theory of unbounded linear operators is developed with the normed linear spaces assumed complete and the operators assumed closed only when needed. The surprising thing is that the proofs are as uncomplicated as the proofs for bounded operators. Thus, the reader who is meeting the theory of linear operators for the first time is not distracted by any of the additional complications which one expects when the operator is not required to be bounded. Chapter III introduces the class of strictly singular operators which includes the class of compact operators. The main reason for considering such a class is to generalize the Riesz-Schauder theory for compact operators. In Chapter V some of the main theorems concerning perturbations of unbounded operators are given and are later applied to ordinary differential operators. In Chapter VI a number of the powerful theorems proved in the earlier chapters are used to examine existence and uniqueness of solutions of certain differential equations. The reader is not required to have any previous knowledge of differential equations. Chapter VII sketches the interplay between functional analysis and "hard" classical analysis in the study of elliptic partial differential equations.

For those unacquainted with the subject matter, examples and motivations for certain definitions and proofs are mentioned in order to give some feeling for what is going on. Simple notation is used so that it is not necessary to refer continually to a collection of symbols in the rear of the book.

In the spring of 1964 I had the great privilege of visiting a number of mathematicians in various parts of the Soviet Union in order to discuss the contents of this book. To Professors I. M. Gelfand and O. Ladyzenskaya goes my gratitude for inviting me to speak about my work at their respective seminars at Moscow State University and the University of Leningrad. I wish to thank Professors M. S. Birman, I. C. Gokhberg, M. G. Krein, A. S. Markus, and M. A. Naimark for their valuable suggestions and comments. I am especially indebted to Professors Gokhberg and Markus for their advice during the three days we spent together in Kishinev.

My gratitude goes to Professors G. Stampacchia and H. G. Tillman for arranging my stay at the Universities of Pisa and Mainz, respectively, where I benefited from their knowledge and experience. Most of the manuscript was written while I was on leave at the University of Pisa.

I wish to express my profound thanks to Professor T. Kato, who was kind enough to show me portions of his manuscript concerned with perturbation theory. Our conversation at Berkeley and our correspondence have been of great help to me.

My appreciation is extended to Professors T. W. Gamelin, G. C. Rota, and R. J. Whitley, who read portions of the manuscript and gave suggestions. Professor Rota was a constant source of encouragement to me in the preparation of this manuscript.

I am especially indebted to Professor R. S. Freeman and my class of 1964–1965 for going through the entire manuscript. Professor Freeman was also kind enough to discuss partial differential equations with me these many months and to offer valuable suggestions.

By writing this book, I have come to realize fully why authors express their gratitude to their typists. It was indeed my good fortune to have Mrs. Ouida Taylor type the manuscript for me. Her accuracy, speed, and artistic layout of each page saved me many months of tedious work.

Finally, my appreciation and thanks go to the Mathematics Division of the Air Force Office of Scientific Research for supporting the major portion of this book under grant number AF OSR 495-64.

Seymour Goldberg

CONTENTS

UNBOUNDED LINEAR OPERATORS

PRELIMINARIES

In a number of sections in this book the standard theorems listed below are used. The proofs of the theorems appear in Dunford and Schwartz [1],† Theorems I.6.9, I.6.15, and IV.6.7 and Lemmas III.3.2, XIV.2.1, XIV.2.2, and XIV.3.3.

0.1 Baire category theorem. *If a complete metric space is the union of a countable number of closed sets, then at least one of the closed sets contains a nonempty open set.*

0.2 Definition. *A set S in a metric space is* ***totally bounded*** *if for every $\varepsilon > 0$ there exists a finite number of spheres, each with center in S and radius ε, which cover S.*

0.3 Theorem. *A set in a complete metric space is totally bounded if and only if its closure is compact.*

0.4 Definition. *Let S be a compact topological space and let $C(S)$ be the metric space of continuous complex-valued functions on S with the metric d given by*

$$d(f, g) = \sup_{x \in S} |f(x) - g(x)|$$

† The number in brackets following a reference refers to a Bibliography entry under that name.

A subset $K \subset C(S)$ is said to be **equicontinuous** *if to every $\varepsilon > 0$ and every $s \in S$ there corresponds an open set $\mathcal{O}_s$ containing s such that for all $f \in K$ and $t \in \mathcal{O}_s$*

$$|f(s) - f(t)| < \varepsilon$$

0.5 Theorem. Ascoli-Arzelà Theorem. *A set $K \subset C(S)$ is compact if and only if it is closed, bounded, and equicontinuous.*

0.6 Hölder's inequality. *Let μ be a measure defined on a σ-ring Σ of subsets of a set Ω. If f and g are μ-measurable complex-valued functions on Ω and if $|f|^p$ and $|g|^{p'}$ are μ-integrable on Ω, where $1/p + 1/p' = 1$, $1 < p < \infty$, then fg is μ-integrable on Ω and*

$$\int_\Omega |fg| \leq \left(\int_\Omega |f|^p\right)^{1/p} \left(\int_\Omega |g|^{p'}\right)^{1/p'}$$

When μ is Lebesgue measure and Σ is the class of Lebesgue measurable subsets of Ω, we write $f \in \mathfrak{L}_p(\Omega)$.

0.7 Definition. *Let Ω be an open set in Euclidean n-space E^n. By $C^k(\Omega)$, $0 \leq k \leq \infty$, we denote the set of all complex-valued functions defined on Ω whose partial derivatives of order $\leq k$ all exist and are continuous. $C_0{}^k$ denotes those functions in $C^k(\Omega)$ which vanish outside a compact set in Ω. The functions are said to have* **compact support** *in Ω. If $\bar{\Omega}$ is the closure of Ω, then $C^k(\bar{\Omega})$ is the set of functions which are restrictions to $\bar{\Omega}$ of those $f \in C^k(\mathcal{O})$, where $\mathcal{O}$ is an open set containing $\bar{\Omega}$. We take $C_0^\infty(\bar{\Omega})$ to be $C_0^\infty(\Omega)$.*

0.8 Theorem. *Let K be a compact subset of E^n and let $\mathcal{O}$ be an open set containing K. Then there exists a function $\varphi \in C^\infty(E^n)$ such that*

i. $0 \leq \varphi(x) \leq 1$, $x \in E^n$
ii. $\varphi(x) = 1$, $x \in K$
iii. $\varphi(x) = 0$, $x \notin \mathcal{O}$

0.9 Theorem. *Let Ω be an open subset of E^n. Given $f \in \mathfrak{L}_p(\Omega)$, $1 \leq p < \infty$, there exists a sequence $\{\varphi_n\}$ in $C_0^\infty(\Omega)$ such that*

$$\int_\Omega |f - \varphi_n| \to 0$$

In terms of the Banach space $\mathfrak{L}_p(\Omega)$, which is discussed in the next chapter, the theorem states that $C_0^\infty(\Omega)$ is dense in $\mathfrak{L}_p(\Omega)$, $1 \leq p < \infty$.

0.10 Theorem. *Let Ω be an open subset of E^n and let f be locally integrable on Ω; that is, f is integrable on every compact subset of Ω. Suppose that for all $\varphi \in C_0^\infty(\Omega)$*

$$\int_\Omega f\varphi = 0$$

Then $f = 0$ a.e.

The proof of the following theorem appears in Loomis [1], page 122.

0.11 Theorem. *Given* $f \in \mathfrak{L}_p(-\infty, \infty)$, $1 \leq p < \infty$, *and* $g \in \mathfrak{L}_1(-\infty, \infty)$, *the* **convolution** $h = f*g$ *of* f *and* g *defined a.e. by*

$$h(x) = \int_{-\infty}^{\infty} f(x - t)g(t)\, dt$$

is in $\mathfrak{L}_p(-\infty, \infty)$ *and*

$$\left(\int_{-\infty}^{\infty} |h(x)|^p\right)^{1/p} \leq \left(\int_{-\infty}^{\infty} |f(x)|^p\right)^{1/p} \int_{-\infty}^{\infty} |g(x)|$$

chapter I

INTRODUCTION TO NORMED LINEAR SPACES

I.1 EXAMPLES OF NORMED LINEAR SPACES

I.1.1 Definition. *Let X be a vector space over the field of real or complex numbers. A* **norm** *on X, denoted by $\| \ \|$, is a real-valued function on X with the following properties:*

i. $\|x\| \geq 0$ *for all* $x \in X$
ii. $x \neq 0$ *implies* $\|x\| \neq 0$
iii. $\|\alpha x\| = |\alpha| \, \|x\|$
iv. $\|x + y\| \leq \|x\| + \|y\|$ *(triangular inequality)*

The vector space X, together with a norm on X, is called a normed linear space. When the scalars over X are the reals, X is called a real normed linear space.

Unless otherwise stated, normed linear spaces are to be considered different from the zero normed linear space.

I.1.2 Definition. *Let X be a normed linear space. If X_1 is a subspace, in the vector space sense, of X and the norm on X_1 is taken to be the restriction of the norm on X, then X_1 is called a subspace of X.*

For A a subset of X, the **span** *of A, written* sp (A), *is the subspace of X consisting of the linear combinations of elements in A. The closure in a topological space of the span of A is denoted by* $\overline{\text{sp}}\ (A)$.

The following examples of normed linear spaces are frequently referred to throughout the text. The vector space operations are the usual ones of pointwise addition and scalar multiplication for functions and component addition and scalar multiplication for n-tuples.

I.1.3 **Example.** *Euclidean* n-space, denoted by E^n, is the normed linear space of n-tuples of real numbers over the reals with norm

$$\|(a_1, a_2, \ldots, a_n)\| = \left(\sum_{i=1}^{n} |a_i|^2\right)^{\frac{1}{2}}$$

Unitary n-space, denoted by U^n, is the normed linear space of n-tuples of complex numbers over the complex numbers with the above norm.

I.1.4 **Example.** For S an arbitrary set, $B(S)$ is the normed linear space of bounded complex-valued functions over the complex numbers with norm

$$\|f\| = \sup_{s \in S} |f(s)|$$

I.1.5 **Example.** For S a compact topological space, $C(S)$ is the subspace of $B(S)$ consisting of the continuous functions.

I.1.6 **Example.** Let $1 \leq p < \infty$ and let μ be a measure defined on a σ-ring Σ of subsets of a set S. Define $\mathfrak{L}_p{}^0(S, \Sigma, \mu)$ to be the class of all those μ-measurable complex-valued functions f for which $|f|^p$ is integrable. Functions f and g in $\mathfrak{L}_p{}^0(S, \Sigma, \mu)$ are said to be equivalent if $f = g$ almost everywhere. $\mathfrak{L}_p(S, \Sigma, \mu)$ will denote the normed linear space of equivalence classes $[f]$ of $f \in \mathfrak{L}_p{}^0(S, \Sigma, \mu)$ with norm given by

$$\|[f]\|_p = \left(\int_S |f|^p \, d\mu\right)^{1/p}$$

Addition and scalar multiplication are defined by

$$\alpha[f] + \beta[g] = [\alpha f + \beta g]$$

As is customary, we shall use f instead of $[f]$ as an element in $\mathfrak{L}_p(S, \Sigma, \mu)$. The proof of the triangular inequality, called Minkowski's inequality, appears in Dunford and Schwartz [1], Lemma II.3.3.

For $p = \infty$, define $\mathfrak{L}_\infty{}^0(S, \Sigma, \mu)$ to be the class of all those μ-measurable complex-valued functions f which are essentially bounded on S; that is, there exists a set $Z \subset S$ such that $\mu(Z) = 0$ and f is bounded on $S \sim Z$, the complement of Z in S. As above, $\mathfrak{L}_\infty(S, \Sigma, \mu)$ will denote the normed linear space of equivalence classes $[f]$ with norm given by

$$\|[f]\|_\infty = \inf_{Z} \sup_{s \in S \sim Z} |f(s)|$$

where Z ranges over the subsets of S of μ-measure zero.

When μ is Lebesgue measure and Σ is the class of Lebesgue-measurable sets, we write $\mathfrak{L}_p(S)$ instead of $\mathfrak{L}_p(S, \Sigma, \mu)$.

I.1.7 Example. In Example I.1.6, let S be the set of positive integers and let Σ be the class of all subsets of S. Define $\mu(E)$ as the number of elements of E and let l_p denote $\mathfrak{L}_p(S, \Sigma, \mu)$, $1 \leq p \leq \infty$. Thus for $1 \leq p < \infty$, l_p is the normed linear space of those sequences $x = (x_1, x_2, \ldots)$, $\sum_{i=1}^{\infty} |x_i|^p < \infty$. The norm is defined by $\|x\|_p = \left(\sum_{i=1}^{\infty} |x_i|^p\right)^{1/p}$. When $p = \infty$, l_∞ is the normed linear space of bounded sequences

$$x = (x_1, x_2, \ldots) \qquad \text{with} \qquad \|x\|_\infty = \sup_{1 \leq i} |x_i|$$

For additional examples of normed linear spaces, we refer the reader to Dunford and Schwartz [1], pages 238–243.

I.2 COMPLETE NORMED LINEAR SPACES

I.2.1 Definition. *Let X be a normed linear space. The metric d induced by the norm is defined by $d(x, y) = \|x - y\|$. For $x \in X$ and M a subset of X, $d(x, M)$ will denote the distance from x to M; that is,*

$$d(x, M) = \inf_{m \in M} \|x - m\|$$

If X is a complete metric space with respect to d, X is called a complete normed linear space or a Banach space. The metric topology on X determined by d is the topology used throughout the book.

I.2.2 Example. $B(S)$ is complete. To see this, suppose $\{f_n\}$ is a Cauchy sequence in $B(S)$. Given $\varepsilon > 0$, there exists an integer N such that for all $s \in S$,

$$(1) \qquad |f_n(s) - f_m(s)| \leq \|f_n - f_m\| \leq \varepsilon \qquad m, n \geq N$$

Hence for each $s \in S$, $\{f_n(s)\}$ is a Cauchy sequence of scalars and therefore converges. Define f on S by $f(s) = \lim_{n \to \infty} f_n(s)$. It is now shown that f is in $B(S)$ and that $f_n \to f$ in $B(S)$. Since $\{f_n\}$ converges pointwise to f, it follows from (1), after fixing $n \geq N$ and letting $m \to \infty$, that for all $s \in S$

$$(2) \qquad |f_n(s) - f(s)| \leq \varepsilon \qquad n \geq N$$

Since f_N is bounded, (2) implies that f is in $B(S)$ and that

$$\|f_n - f\| = \sup_{s \in S} |f_n(s) - f(s)| \leq \varepsilon \qquad n \geq N$$

Thus $f_n \to f$ in $B(S)$.

I.2.3 Example. To prove that $C(S)$ is complete, it suffices to prove that $C(S)$ is closed in $B(S)$, since $C(S)$ is a subspace of Banach space $B(S)$. Suppose $f_n \to f$ in $B(S)$, where $\{f_n\}$ is a sequence in $C(S)$. We need only show that f is continuous on S. Given $\varepsilon > 0$, there exists an integer N such that

$$(1) \qquad |f(s) - f_N(s)| \leq \|f - f_N\| \leq \frac{\varepsilon}{3} \qquad s \in S$$

Since f_N is continuous on S, there exists, for each $s \in S$, an open set O_s containing s such that

$$(2) \qquad |f_N(s) - f_N(t)| \leq \frac{\varepsilon}{3} \qquad t \in O_s$$

Thus for $t \in O_s$, (1) and (2) imply

$$\begin{aligned} |f(s) - f(t)| &\leq |f(s) - f_N(s)| + |f_N(s) - f_N(t)| + |f_N(t) - f(t)| \\ &\leq \frac{\varepsilon}{3} + \frac{\varepsilon}{3} + \frac{\varepsilon}{3} \end{aligned}$$

Hence f is in $C(S)$.

The proof of the completeness of $\mathfrak{L}_p(S, \Sigma, \mu)$ appears in Dunford and Schwartz [1].

I.2.4 Definition. *As in the definition of the sum of an infinite series of numbers, we say that an infinite series* $\sum_{i=1}^{\infty} x_i$ *of elements* x_i *in normed linear space* X *converges in* X *if there exists an* $x \in X$ *such that* $s_n = \sum_{i=1}^{n} x_i$ *converges to* x*. We write* $x = \sum_{i=1}^{\infty} x_i$.

The series ***converges absolutely*** *if* $\sum_{i=1}^{\infty} \|x_i\| < \infty$.

I.2.5 Theorem. *A normed linear space* X *is complete if and only if every series in* X *which converges absolutely also converges in* X.

Proof. Suppose X is complete and $\sum_{i=1}^{\infty} \|x_i\| < \infty$, $x_i \in X$. For

$$s_n = \sum_{i=1}^{n} x_i,$$

$$\|s_{n+k} - s_n\| = \left\| \sum_{i=n+1}^{n+k} x_i \right\| \leq \sum_{i=n+1}^{n+k} \|x_i\| \to 0 \qquad \text{as } n \to \infty$$

Thus $\{s_n\}$ is a Cauchy sequence and therefore converges in Banach space X. Conversely, assume absolute convergence implies convergence. Let $\{y_n\}$ be a Cauchy sequence in X. There exists a subsequence $\{x_n\}$ of $\{y_n\}$ such that

$$\|x_{k+1} - x_k\| \leq 2^{-k} \qquad 1 \leq k$$

Now

$$x_k = x_1 + (x_2 - x_1) + \cdots + (x_k - x_{k-1}) \qquad 2 \leq k$$

and

$$\sum_{i=1}^{\infty} \|x_{i+1} - x_i\| \leq \sum_{i=1}^{\infty} 2^{-i} = 1$$

Thus, by hypothesis, the sequence of partial sums $x_k - x_1$ converges in X, or equivalently, the sequence $\{x_k\}$ converges to some $x \in X$. Since $\{x_k\}$ is a subsequence of the Cauchy sequence $\{y_k\}$, it is easy to see that $\{y_k\}$ also converges to x. Hence X is complete.

In linear algebra one encounters the concept of a quotient space X/M, where M is a subspace of vector space X. When X is a normed linear space and M is a closed subspace of X, a norm is put on X/M so that certain topological properties of operators defined on X are shared by corresponding operators on X/M. Quotient spaces which are complete play a useful role, particularly in Chaps. II, IV, and V.

I.2.6 Definition. *Let M be a closed subspace of normed linear space X. Define an equivalence relation R on X by xRy if $x - y$ is in M. Let X/M denote the corresponding set of equivalence classes and let $[x]$, called a coset, denote the set of elements equivalent to x. Thus*

$$[x] = \{x + m \mid m \in M\} = x + M$$

Vector addition and scalar multiplication on X/M are defined by

$$[x] + [y] = [x + y]$$

$$\alpha[x] = [\alpha x]$$

Define a norm on X/M by

$$\|[x]\| = d(x, M)$$

where $d(x, M)$ is the distance from x to M. It is left to the reader to verify that X/M is a normed linear space. M is required to be closed in order that $\|[x]\| = 0$ implies $[x] = [0]$.

Geometrically, if X is the plane and M is a line through the origin, then X/M is the space whose elements are M together with the lines parallel to M. The norm of $[x]$ is the distance between the line containing x and the line M.

I.2.7 Remarks

i. Since any $y \in [x]$ is of the form $x - m$, $m \in M$, it follows that $\|[x]\| = \inf_{y \in [x]} \|y\|$.

ii. If $[x]$ and $[z]$ are such that $\|[x] - [z]\| < \varepsilon$, there exists a $v \in [z]$ such that $\|x - v\| < \varepsilon$.

I.2.8 Theorem. *If X is a Banach space and M is a closed subspace of X, then X/M is a Banach space.*

Thinking of cosets as being lines, the proof proceeds by considering a given Cauchy sequence of parallel lines (lines "crowded" together) and choosing points, one on each line, which are crowded together in the sense of being a Cauchy sequence. The Cauchy sequence of points converges to a point x, and the sequence of lines containing the points converges to a line containing x.

Proof. Let $\{[x_n]\}$ be a Cauchy sequence in X/M. There exists a subsequence $\{[y_n]\}$ of $\{[x_n]\}$ such that

$$\|[y_{n+1}] - [y_n]\| < 2^{-n} \qquad 1 \leq n$$

By Remark (*ii*) of I.2.7, we may choose $v_n \in [y_n]$ so that

$$\|v_{n+1} - v_n\| < 2^{-n}$$

The sequence $\{v_n\}$ is a Cauchy sequence, since

$$\|v_{n+i} - v_n\| \leq \|v_{n+1} - v_n\| + \|v_{n+2} - v_{n+1}\| + \cdots + \|v_{n+i} - v_{n+i-1}\|$$
$$\leq \sum_{k=0}^{\infty} 2^{-n-k} = 2^{-n+1} \to 0 \qquad \text{as } n \to \infty$$

The completeness of X assures the existence of a $v \in X$ such that $v_n \to v$. By Remark (*i*) of I.2.7,

$$\|[v_n] - [v]\| = \|[v_n - v]\| \leq \|v_n - v\|$$

Thus $[v_n]$ converges to $[v]$ in X/M. Since $\{[v_n]\} = \{[y_n]\}$ is a subsequence of the Cauchy sequence $\{[x_n]\}$, $\{[x_n]\}$ also converges to $[v]$. Hence X/M is complete. A simpler proof makes use of Theorem I.2.5.

We shall show in Theorem I.6.16 that every normed linear space is a dense subspace of a Banach space.

I.3 BOUNDED LINEAR OPERATORS

I.3.1 Definition. *Let X and Y be vector spaces over the same space of scalars. An operator T with domain X and range in Y is called linear if for all x and z in X and all scalars α and β,*

$$T(\alpha x + \beta z) = \alpha Tx + \beta Tz$$

For the remainder of the book the following notations are used.

$\mathfrak{D}(T)$ *denotes the domain of T.*
$\mathfrak{R}(T)$ *denotes the range of T.*
$\mathfrak{N}(T)$ *denotes the subspace $\{x \mid Tx = 0\}$. $\mathfrak{N}(T)$ is called the null manifold or kernel of T.*

T is called 1-1 if distinct elements in $\mathfrak{D}(T)$ are mapped into distinct elements in $\mathfrak{R}(T)$. Since a linear operator T has the property that $T0 = 0$, T is 1-1 if and only if $\mathfrak{N}(T) = (0)$.

I.3.2 Theorem. *Let X and Y be normed linear spaces and let T be a linear operator with domain X and range in Y. The following statements are equivalent.*

i. T is continuous at a point.
ii. T is uniformly continuous on X.
iii. T is bounded; i.e., there exists a number M such that for all $x \in X$

$$\|Tx\| \leq M\|x\|$$

Proof. *(i) implies (ii).* Suppose T is continuous at x_0. Given $\varepsilon > 0$, there exists a $\delta = \delta(\varepsilon)$ such that

$$\|Tx - Tx_0\| < \varepsilon \qquad \|x - x_0\| < \delta \tag{1}$$

Let u be any point in X. Then for $\|u - v\| < \delta$ it follows from (1) and the additivity of T that

$$\|Tv - Tu\| = \|Tx_0 - T(x_0 + u - v)\| < \varepsilon$$

Hence *(ii)* is proved.

(ii) implies (iii). The continuity of T at 0 implies the existence of a $\delta > 0$ such that

$$\|Tz\| = \|Tz - T0\| \leq 1 \qquad \|z\| \leq \delta$$

Thus, for $x \neq 0$ in X, $\delta = \|\delta x/\|x\|\|$, and therefore

$$1 \geq \left\| T\left(\frac{\delta x}{\|x\|}\right) \right\| = \frac{\delta\|Tx\|}{\|x\|}$$

Hence $\|Tx\| \leq \delta^{-1}\|x\|$ for all $x \in X$.

(iii) implies (i). The inequality $\|Tx - Tz\| \leq M\|x - z\|$ implies (*i*).

I.3.3 Definition. *Let X and Y be normed linear spaces. Define the norm on the vector space of bounded linear operators which map X into Y by*

$$\|T\| = \sup_{\|x\|=1} \|Tx\| = \sup_{x \neq 0} \frac{\|Tx\|}{\|x\|}$$

It is easy to verify that this is indeed a norm. $[X, Y]$ will denote the normed linear space of bounded linear operators with the above norm. For $X = Y$, $[X]$ will be used instead of $[X, X]$.

I.3.4 Example. Take $X = Y = C(I)$, where I is a compact interval on the line. Let k be a real-valued function which is continuous on the rectangle $I \times I$ and let K be the linear map from X into Y defined by

$$(Kx)(t) = \int_I k(s, t)x(s)\, ds$$

From the continuity of k it follows that $\mathcal{R}(K) \subset Y$. We shall determine $\|K\|$. Now

$$|(Kx)(t)| \leq \|x\| \int_I |k(s, t)|\, ds$$

Thus

$$\|K\| \leq \max_{t \in I} \int_I |k(s, t)|\, ds \tag{1}$$

Since

$$y(t) = \int_I |k(s, t)|\, ds$$

is continuous on I, y attains its maximum at some point $t_0 \in I$. Define

$$x(s) = \begin{cases} \dfrac{k(s, t_0)}{|k(s, t_0)|} & \text{when } k(s, t_0) \neq 0 \\ 0 & \text{otherwise} \end{cases}$$

Then x is integrable on I since it is bounded and measurable. It follows from Theorem 0.9 that there exists a sequence $\{x_n\}$ in $C(I)$ such that

$\|x_n\| \leq 1$ and $\{x_n\}$ converges to x in $\mathfrak{L}_1(I)$. Therefore

$$\|K\| \geq \|Kx_n\| \geq (Kx_n)(t_0) \to \int_I k(s, t_0)x(s)\, ds$$

Hence

$$(2) \qquad \|K\| \geq \int_I k(s, t_0)x(s)\, ds = \int_I |k(s, t_0)|\, ds = \max_{t \epsilon I} \int_I |k(s, t)|\, ds$$

Thus, by (1) and (2),

$$\|K\| = \max_{t \epsilon I} \int_I |k(s, t)|\, ds$$

I.3.5 Theorem. *If X is a normed linear space and Y is a Banach space, then $[X, Y]$ is a Banach space.*

Proof. The proof is essentially the same as the proof in I.2.2, which shows that $B(S)$ is complete.

The converse to the theorem is proved in Corollary I.5.8.

I.3.6 Definition. *Let T be a* 1-1 *linear operator with $\mathfrak{D}(T) \subset X$ and $\mathfrak{R}(T) \subset Y$, X and Y normed linear spaces. The inverse of T, written T^{-1}, is the map from subspace $\mathfrak{R}(T)$ into X given by $T^{-1}(Tx) = x$. It is clear that T^{-1} is linear.*

For a function f which is not necessarily 1-1, *$f^{-1}(B)$ will be used to denote the set $\{x \mid f(x) \epsilon B\}$.*

I.3.7 Theorem. *Let T be a linear map from normed linear space X into normed linear space Y. T^{-1} exists and is continuous if and only if there exists an $m > 0$ such that*

$$\|Tx\| \geq m\|x\| \qquad x \epsilon X$$

Proof. Suppose $\|Tx\| \geq m\|x\|$ for all $x \epsilon X$. Then $x \neq 0$ implies $Tx \neq 0$. Hence T is 1-1. Since

$$\|T^{-1}Tx\| = \|x\| \leq m^{-1}\|Tx\|$$

T^{-1} is bounded and therefore continuous. On the other hand, if T^{-1} is continuous, then

$$\|x\| = \|T^{-1}Tx\| \leq \|T^{-1}\|\, \|Tx\| \qquad x \epsilon X$$

The theorem follows upon taking $m = 1/\|T^{-1}\|$.

I.4 FINITE–DIMENSIONAL NORMED LINEAR SPACES

I.4.1 ***Definition.*** *A linear operator mapping a normed linear space into a normed linear space is called an* **isomorphism** *if it is continuous and has a continuous inverse.*

Normed linear spaces X and Y are said to be **isomorphic** *if there exists an isomorphism mapping X onto Y.*

I.4.2 ***Theorem.*** *Every n-dimensional normed linear space with complex (real) scalars is isomorphic to unitary (Euclidean) n-space.*

Proof. Assume, for definiteness, that X is an n-dimensional normed linear space over the complex numbers. Let $x_1, x_2, \ldots, x_n$ be a basis for X. Define the linear map T from U^n onto X by

$$T(\alpha_1, \alpha_2, \ldots, \alpha_n) = \sum_{i=1}^{n} \alpha_i x_i$$

It is easily seen that T is bounded and therefore continuous. It remains to prove that T has a bounded inverse. Define a real-valued function f on U^n by $f(\alpha) = \|T\alpha\|$. The continuity of both T and the norm imply the continuity of f. Since the sphere $S = \{\alpha \mid \alpha \in U^n, \|\alpha\| = 1\}$ is compact, f assumes a minimum at some point $\gamma \in S$. Moreover, $f(\gamma) > 0$, otherwise

$$0 = T\gamma = \sum_{i=1}^{n} \gamma_i x_i \qquad \gamma = (\gamma_1, \gamma_2, \ldots, \gamma_n)$$

which is impossible since $\gamma \neq 0$ and the set of x_i is linearly independent. For $\alpha \neq 0$ in U^n, $\alpha/\|\alpha\|$ is in S and $\|T(\alpha/\|\alpha\|)\| \geq \|T\gamma\|$. Thus, for all $\alpha \in U^n$,

$$\|T\alpha\| \geq m\|\alpha\| \qquad m = \|T\gamma\| > 0$$

which implies that T has a bounded inverse.

Since isomorphisms take closed bounded sets onto closed bounded sets and since a set in U^n or E^n is compact if and only if it is closed and bounded, the following corollary is obtained.

I.4.3 ***Corollary.*** *A closed bounded set in a finite-dimensional normed linear space is compact.*

I.4.4 ***Definition.*** *For X a normed linear space and r a positive number, the* **r-ball** *of X is the set* $\{x \mid \|x\| \leq r, x \in X\}$.

The **r-sphere** *of X is the set* $\{x \mid \|x\| = r, x \in X\}$.

From Corollary I.4.3 we know that the 1-sphere of a finite-dimensional normed linear space is compact. Theorem I.4.6, which is due to F. Riesz, shows that this property of the 1-sphere characterizes the finite-dimensional normed linear spaces.

I.4.5 Lemma. *If N is a finite-dimensional proper subspace of normed linear space X, there exists an element in the* 1*-sphere of X whose distance from N is* 1.

Proof. Choose z to be a point in X but not in N. There exists a sequence $\{n_k\}$ in N such that $\|z - n_k\| \to d(z, N)$. Since N is finite-dimensional and $\{n_k\}$ is bounded, Corollary I.4.3 implies the existence of a subsequence $\{v_k\}$ of $\{n_k\}$ which converges to some $n \in N$. Hence

$$\|z - n\| = \lim_{k \to \infty} \|z - v_k\| = d(z, N) = d(z - n, N)$$

Since $z - n \neq 0$, it follows that

$$1 = \frac{\|z - n\|}{\|z - n\|} = \frac{d(z - n, N)}{\|z - n\|} = d\left(\frac{z - n}{\|z - n\|}, N\right)$$

I.4.6 Theorem. *If X is a normed linear space whose* 1*-sphere is totally bounded, then X is finite-dimensional.*

Proof. By hypothesis, there exist points $x_1, x_2, \ldots, x_k$ in the 1-sphere S of X such that given $x \in S$, there is an x_i such that

$$\|x - x_i\| < 1 \tag{1}$$

Assert that the finite-dimensional space N spanned by $x_1, x_2, \ldots, x_k$ is X. Suppose the assertion is false. Then Lemma I.4.5 assures the existence of an $x \in S$ whose distance from N is 1. But this contradicts (1).

Lemma I.4.5 need not hold for infinite-dimensional N. See, for example, Taylor [1], page 97. However, we prove the following result.

I.4.7 Riesz's lemma. *Let M be a subspace of normed linear space X which is not dense in X. There exists a sequence $\{x_n\}$ in the* 1*-sphere of X such that $d(x_n, M) \to 1$.*

The proof is motivated by the following geometric consideration. Let X be the plane and let M be the horizontal axis. Take x to be any vector not in M and draw some vectors $x - m$, $m \in M$. Upon normalizing the vectors $x - m$ so that they lie on the unit circle, it is seen that for certain m the distance from $(x - m)/\|x - m\|$ to M exceeds $1 - \varepsilon$, $\varepsilon > 0$.

Proof. Let S be the 1-sphere of X. Since 0 is in M, the distance from any element in S to M does not exceed 1. Thus, to prove the lemma, it suffices to prove that for $1 > \varepsilon > 0$, there exists an element in S whose distance from M is not less than $1 - \varepsilon$. By hypothesis, there exists an $x \in X$ such that $d(x, M) > 0$. We show the existence of an $m \in M$ which satisfies

$$d\left(\frac{x - m}{\|x - m\|}, M\right) \geq 1 - \varepsilon$$

thereby proving the lemma. Since for any $m \in M$,

$$d\left(\frac{x - m}{\|x - m\|}, M\right) = \frac{d(x - m, M)}{\|x - m\|} = \frac{d(x, M)}{\|x - m\|}$$

the problem reduces to finding an m so that

$$\frac{d(x, M)}{\|x - m\|} \geq 1 - \varepsilon$$

or, equivalently,

$$\frac{d(x, M)}{1 - \varepsilon} \geq \|x - m\| \tag{1}$$

Since $d(x, M)/(1 - \varepsilon) > d(x, M)$, it follows from the definition of $d(x, M)$ that there exists an $m \in M$ for which (1) holds.

Riesz's lemma, in place of Lemma I.4.5, is usually used to prove Theorem I.4.6.

I.4.8 Lemma. *If normed linear space Y is isomorphic to a Banach space, then Y is also a Banach space.*

Proof. Let T be an isomorphism from Banach space X onto Y. Suppose $\{y_n\}$ is a Cauchy sequence in Y. Since T^{-1} is bounded,

$$\|T^{-1}y_n - T^{-1}y_m\| \leq \|T^{-1}\| \, \|y_n - y_m\|$$

Hence $\{T^{-1}y_n\}$ is also a Cauchy sequence and therefore converges to some $x \in X$. By the continuity of T, $y_n = TT^{-1}y_n \rightarrow Tx$. Therefore Y is complete.

I.4.9 Corollary. *Every finite-dimensional normed linear space is complete.*

Proof. Lemma I.4.8 and Theorem I.4.2.

Since a subspace which is complete is also closed, the next corollary is an immediate consequence of Corollary I.4.9.

I.4.10 Corollary. *A finite-dimensional subspace of a normed linear space is closed.*

I.4.11 Definition. *For A and B both subsets of a vector space the sum of A and B, denoted by $A + B$, is the set $\{a + b \mid a \in A,\ b \in B\}$. So as not to confuse $A + B$ with the union of A and B, $A \cup B$ will always be used to denote the union of the sets.*

I.4.12 Theorem. *The sum of two closed subspaces of a normed linear space is closed whenever one of the subspaces is finite-dimensional.*

Proof. Let M and N be a closed subspace and a finite-dimensional subspace, respectively, of normed linear space X. Define the natural linear map A from X onto X/M by $Ax = [x]$. Since $\|Ax\| = \|[x]\| \leq \|x\|$, A is continuous. Moreover, the linearity of A and the finite-dimensionality of N imply the finite-dimensionality of AN. Hence, by Corollary I.4.10, AN is closed and therefore $A^{-1}AN = M + N$ is closed. (Note that A^{-1} is used in the set theoretic sense.)

The following example shows that the sum of two closed subspaces of a Banach space need not be closed. In fact, it shows that the sum of two closed subspaces can be a proper dense subspace.

I.4.13 Example. Let $Z = l_2 \times l_2$ with norm $\|(x, y)\| = \|x\| + \|y\|$ and let M be the closed subspace $l_2 \times \{0\}$ of Z. Choose T to be any bounded, 1-1 linear map from l_2 onto a proper dense subspace of l_2; for example, $T(\{\alpha_k\}) = \{\alpha_k/k\} \in l_2$. Let $N = \{(x, Tx) \mid x \in l_2\}$. Clearly, N is a closed subspace of Z. Since T is 1-1 and $\mathcal{R}(T) \neq l_2$, $M \cap N = (0, 0)$ and $M + N \neq Z$. However, $M + N$ is dense in Z; for given $(x, y) \in Z$, there exists a sequence $\{Tx_n\}$ which converges to y, and therefore $(x - x_n, 0) + (x_n, Tx_n) \to (x, y)$.

I.5 HAHN–BANACH EXTENSION THEOREM

I.5.1 Definition. *A functional on a vector space V is a map from V to the scalars. The* **conjugate** *X' of a normed linear space X is the Banach space of bounded linear functionals on X; that is, $X' = [X, Y]$, where Y is the Banach space of scalars with absolute value taken as norm. Note that X' is complete by Theorem* I.3.5.

We now come to one of the most fundamental theorems in functional analysis.

I.5.2 Hahn-Banach extension theorem. *Suppose X is a vector space over the real or complex numbers. Let M be a subspace of X and let p be a real-valued function on X with the following properties.*

i. $p(x + y) \leq p(x) + p(y)$
ii. $p(\alpha x) = |\alpha| p(x)$

If f is a linear functional on M such that

$$|f(m)| \leq p(m) \qquad m \in M$$

then there exists a linear functional F which is an extension of f to all of X such that

$$|F(x)| \leq p(x) \qquad x \in X$$

Proof. Let $\mathcal{P}$ be the set of all linear extensions g of f with domain in X such that $|g(x)| \leq p(x)$ for all $x \in \mathfrak{D}(g)$. $\mathcal{P}$ is not empty, since $f \in \mathcal{P}$. Partially order $\mathcal{P}$ by having $g \leq k$ mean that k is a linear extension of g. Using Zorn's lemma, it will be shown that $\mathcal{P}$ contains a functional which is defined on all of X, thereby proving the theorem. Let $\mathfrak{J}$ be a totally ordered subset of $\mathcal{P}$. Define H by

$$\mathfrak{D}(H) = \bigcup_{g \in \mathfrak{J}} \mathfrak{D}(g)$$

$$H(x) = g(x) \qquad \text{for any } g \in \mathfrak{J} \text{ with } x \in \mathfrak{D}(g)$$

Since $\mathfrak{J}$ is totally ordered, it follows that H is in $\mathcal{P}$. Obviously, $H \geq g$ for all $g \in \mathfrak{J}$. Hence, by Zorn's lemma, $\mathcal{P}$ contains a maximal element F. It remains to prove that $\mathfrak{D}(F) = X$. Assume there exists an $x \in X$ which is not in $\mathfrak{D}(F)$. Let M_1 be the subspace spanned by x and the elements in $\mathfrak{D}(F)$. $\mathfrak{D}(F)$ is a proper subspace of M_1, and each element in M_1 has a unique representation of the form $\alpha x + m$, $m \in \mathfrak{D}(F)$. A $G \in \mathcal{P}$ will now be constructed with domain M_1, thereby contradicting the fact that F is a maximal element in $\mathcal{P}$. Thus G is to be defined by

$$G(\alpha x + m) = \alpha G(x) + G(m) = \alpha G(x) + F(m) \qquad m \in \mathfrak{D} = \mathfrak{D}(F) \tag{1}$$

with $G(x)$ chosen so that

$$|G(\alpha x + m)| \leq p(\alpha x + m) \qquad m \in \mathfrak{D} \tag{2}$$

The proof of the theorem is now reduced to showing that a number, which we call $G(x)$, exists such that (2) holds.

We first suppose that X is a real vector space. Note that (2) is equivalent to

$$G(\alpha x + m) \leq p(\alpha x + m) \qquad m \in \mathfrak{D} \tag{3}$$

for if (3), then

$$-G(\alpha x + m) = G(-\alpha x - m) \leq p(-\alpha x - m) = p(\alpha x + m)$$

As particular cases of (3), we want $G(x)$ to be defined so that

$$G(x + z) \leq p(x + z) \qquad z \in \mathfrak{D} \tag{4}$$

and

$$-G(x + y) = G(-x - y) \leq p(-x - y) \qquad y \in \mathfrak{D} \tag{5}$$

It turns out, as will be shown presently, that the validity of (4) and (5) is enough to guarantee (3). From (1) we see that (4) and (5) are equivalent to

$$G(x) \leq p(x + z) - F(z) \qquad z \in \mathfrak{D} \tag{6}$$

and

$$G(x) \geq -p(-x - y) - F(y) \qquad y \in \mathfrak{D} \tag{7}$$

respectively. Now, the right side of (6) dominates the right side of (7) for all z and $y \in \mathfrak{D}$, since

$$p(x + z) - F(z) + p(-x - y) + F(y) \geq p(z - y) - F(z - y) \geq 0$$

Hence, defining $G(x) = \inf \{p(x + z) - F(z) \mid z \in \mathfrak{D}\}$, (6) and (7) and therefore (4) and (5) hold. To show that (4) together with (5) implies (3), three cases are considered.

i. For $\alpha > 0$, (4) implies

$$G(\alpha x + m) = \alpha G\left(x + \frac{m}{\alpha}\right) \leq \alpha p\left(x + \frac{m}{\alpha}\right) = p(\alpha x + m) \qquad m \in \mathfrak{D}$$

ii. For $\alpha < 0$, (5) implies

$$G(\alpha x + m) = -\alpha G\left(-x - \frac{m}{\alpha}\right) \leq -\alpha p\left(-x - \frac{m}{\alpha}\right) = p(\alpha x + m) \qquad m \in \mathfrak{D}$$

iii. For $\alpha = 0$

$$G(m) = F(m) \leq p(m) \qquad m \in \mathfrak{D}$$

Thus the theorem for real vector spaces is proved.

Suppose X has complex scalars. The following proof is due to Bohnenblust and Sobczyk. The idea of the proof is to reduce the complex case to the real case and apply the above result. Write

$$f(m) = \operatorname{Re} f(m) + i \operatorname{Im} f(m) \qquad m \in M$$

where $\operatorname{Re} f(m)$ and $\operatorname{Im} f(m)$ denote the real and imaginary parts, respectively, of $f(m)$. Since $f(im) = if(m)$, we see that

$$\operatorname{Re} f(im) + i \operatorname{Im} f(im) = -\operatorname{Im} f(m) + i \operatorname{Re} f(m)$$

Thus $\operatorname{Im} f(m) = -\operatorname{Re} f(im)$ and

$$f(m) = \operatorname{Re} f(m) - i \operatorname{Re} f(im) \tag{8}$$

Let X_r be X considered as a vector space over the reals. As sets, $X = X_r$. Now $\operatorname{Re} f$ is a linear functional on M considered as a subspace of X_r and

$$|\operatorname{Re} f(m)| \leq |f(m)| \leq p(m) \qquad m \in M$$

Hence, by what has already been proved, there exists a linear extension G of $\operatorname{Re} f$ to all of X_r such that

$$|G(x)| \leq p(x) \qquad x \in X_r$$

In view of Eq. (8), define F on X by

$$F(x) = G(x) - iG(ix) \qquad x \in X$$

A simple calculation shows that F is a linear extension of f to all of X. Given $x \in X$, write $F(x)$ in polar form $F(x) = |F(x)|e^{i\theta}$. Then

$$|F(x)| = F(e^{-i\theta}x) = \operatorname{Re} F(e^{-i\theta}x) = G(e^{-i\theta}x) \leq p(e^{-i\theta}x) = p(x)$$

Thus the theorem is proved.

I.5.3 Remark. In the proof of the above theorem, the following result has been shown.

Suppose X is a vector space over the reals. Let M be a subspace of X and let p be a real-valued function on X with the following properties.

$$p(x + y) \leq p(x) + p(y)$$

$$p(\alpha x) = \alpha p(x) \qquad \alpha \geq 0$$

If f is a linear functional on M such that

$$f(m) \leq p(m) \qquad m \in M$$

then there exists a linear functional F which is an extension of f to all of X such that

$$F(x) \leq p(x) \qquad x \in X$$

I.5.4 Corollary. *Let m' be a continuous linear functional on a subspace M of normed linear space X. There exists an $x' \in X'$ such that $x' = m'$ on M and $\|x'\| = \|m'\|$.*

Proof. Define p on X by $p(x) = \|m'\| \, \|x\|$. Since p satisfies (*i*) and (*ii*) of Theorem I.5.2 and $|m'm| \leq \|m'\| \, \|m\| = p(m)$, $m \in M$, there exists a linear functional x' on X such that $x' = m'$ on M and

$$|x'x| \leq p(x) = \|m'\| \, \|x\| \qquad x \in X$$

Thus x' is in X' and $\|x'\| \leq \|m'\|$. On the other hand,

$$\|x'\| = \sup_{\substack{\|x\|=1 \\ x \in X}} |x'x| \geq \sup_{\substack{\|x\|=1 \\ x \in M}} |x'x| = \sup_{\substack{\|x\|=1 \\ x \in M}} |m'x| = \|m'\|$$

Hence $\|x'\| = \|m'\|$.

I.5.5 Corollary. *Let M be a subspace of normed linear space X. Given $x \in X$ with $d = d(x, M) > 0$, there exists an $x' \in X'$ such that*

$$\|x'\| = 1 \qquad x'M = 0 \qquad \text{and} \qquad x'x = d(x, M)$$

Proof. Let M_1 be the subspace spanned by x and the elements of M. Define linear functional v' on M_1 by

$$v'(\alpha x + m) = \alpha d \qquad m \in M$$

Then $v'M = 0$ and $v'x = d$. We assert that v' is in M_1' with $\|v'\| = 1$. For $\alpha \neq 0$ and $m \in M$,

$$\|\alpha x + m\| = |\alpha| \left\| x + \frac{m}{\alpha} \right\| \geq |\alpha| \, d$$

Thus for all α,

$$|v'(\alpha x + m)| = |\alpha|\, d \leq \|\alpha x + m\| \qquad m \in M$$

Hence $v' \in M_1'$ and $\|v'\| \leq 1$. There exists a sequence $\{m_k\}$ in M such that $\|x - m_k\| \to d$. Since

$$d = v'(x - m_k) \leq \|v'\|\, \|x - m_k\| \to \|v'\|\, d$$

it follows that $\|v'\| \geq 1$. Thus $\|v'\| = 1$. The corollary follows upon taking $x' \in X'$ to be an extension of v' so that $\|x'\| = \|v'\| = 1$.

I.5.6 Corollary. *Given $x \in X$, there exists an $x' \in X'$ such that $\|x'\| = 1$ and $x'x = \|x\|$. In particular, if $x \neq y$, there exists an $x' \in X'$ such that $0 \neq \|x - y\| = x'x - x'y$.*

Proof. Take $M = (0)$ in Corollary I.5.5.

I.5.7 Corollary. *For any x in normed linear space X,*

$$\|x\| = \sup_{\substack{\|x'\|=1 \\ x' \in X'}} |x'x|$$

Proof. For x' in the 1-sphere of X'

$$|x'x| \leq \|x'\|\, \|x\| \leq \|x\| \tag{1}$$

By Corollary I.5.6, there exists a z' in the 1-sphere of X' such that

$$z'x = \|x\| \tag{2}$$

The corollary follows from (1) and (2).

As a simple application of Corollary I.5.6, we prove the converse to Theorem I.3.4.

I.5.8 Corollary. *Let X and Y be normed linear spaces. If $[X, Y]$ is complete, then Y is complete.*

Proof. Let $\{y_n\}$ be a Cauchy sequence in Y. Choose $x_0 \in X$ such that $\|x_0\| = 1$. (Recall that $X \neq (0)$ unless stated otherwise.) There exists an $x' \in X'$ such that $x'x_0 = \|x_0\| = 1$. Define $T_n \in [X, Y]$ by

$$T_n x = x'(x) y_n$$

Now

$$\|(T_n - T_m)x\| = |x'x|\, \|y_n - y_m\| \leq \|x'\|\, \|y_n - y_m\|\, \|x\| \qquad x \in X$$

Hence $\|T_n - T_m\| \leq \|x'\| \, \|y_n - y_m\|$ which implies that $\{T_n\}$ is a Cauchy sequence in $[X, Y]$. Thus, by hypothesis, $\{T_n\}$ converges in $[X, Y]$ to some T. Since

$$\|y_n - Tx_0\| = \|T_n x_0 - Tx_0\| \leq \|T_n - T\| \, \|x_0\|$$

$\{y_n\}$ converges to Tx_0 and therefore Y is complete.

The next theorem is very useful and is needed in the proof of Theorem II.4.3.

I.5.9 ***Definition.*** *A subset K of a vector space over the real or complex numbers is called* ***convex*** *if for every x and y in K, the set*

$$\{ax + (1 - a)y \mid 0 \leq a \leq 1\}$$

is contained in K.

I.5.10 ***Theorem.*** *Let K be a closed convex subset of normed linear space X. Given $x \in X$ but not in K, there exists an $f \neq 0 \in X'$ such that*

$$\operatorname{Re} f(x) \geq \operatorname{Re} f(k) \qquad k \in K$$

The conclusion in the above theorem has the following geometric interpretation when X is the plane. Consider $\mathfrak{N}(f)$ as a line through the origin. Choose x_0 so that $f(x_0) = 1$. Now, x lies on the line $f(x)x_0 + \mathfrak{N}(f)$ and each $k \in K$ lies on the line $f(k)x_0 + \mathfrak{N}(f)$. Since

$$f(x) \geq f(k) \qquad k \in K$$

it follows that K lies on one side of the line $f(x)x_0 + \mathfrak{N}(f)$ containing x.

The proof of the theorem depends on the following lemma.

I.5.11 ***Definition.*** *Let 0 be an interior point of a convex subset K of normed linear space X. For each $x \in X$ let*

$$A(x) = \{a \mid a > 0, x \in aK\}$$

where $aK = \{ak \mid k \in K\}$. Define the functional p on X by

$$p(x) = \inf A(x)$$

We shall call p the Minkowski functional of K. Since 0 is an interior point of K, $A(x) \neq \phi$ and $0 \leq p(x) < \infty$.

I.5.12 Lemma. *Let K and p be as in the above definition. Then for x and y in X,*

i. $p(\alpha x) = \alpha p(x),\ \alpha \geq 0$
ii. $p(x + y) \leq p(x) + p(y)$
iii. $p(z) \geq 1$ *for all* $z \notin K$

Proof of (i). It is clear that $p(0) = 0$. Suppose $\alpha > 0$. Given $a \in A(\alpha x)$, x is in $\alpha^{-1}aK$ and therefore

$$p(x) \leq \inf_{a \in A(\alpha x)} \alpha^{-1}a = \alpha^{-1}p(\alpha x)$$

Thus $\alpha p(x) \leq p(\alpha x)$, $\alpha \geq 0$. This result implies

$$p(x) = p(\alpha^{-1}\alpha x) \geq \alpha^{-1}p(\alpha x) \qquad \alpha > 0$$

or $\alpha p(x) \geq p(\alpha x)$. Hence $p(\alpha x) = \alpha p(x)$, $\alpha \geq 0$.

Proof of (ii). We first observe that if a and b are nonnegative real numbers, then $(a + b)K = aK + bK$. Indeed, if x and y are in K, then the convexity of K implies

$$\frac{a}{a+b}x + \frac{b}{a+b}y \in K \qquad a + b \neq 0$$

Thus $ax + by \in (a + b)K$. Since x and y are arbitrary in K, $aK + bK \subset (a + b)K$. Obviously, $(a + b)K \subset aK + bK$. Hence

$$(a + b)K = aK + bK$$

To conclude the proof of *(ii)*, we suppose $a \in A(x)$ and $b \in A(y)$. Then $x + y \in aK + bK = (a + b)K$. Hence $p(x + y) \leq a + b$. Since a and b are arbitrary in $A(x)$ and $A(y)$, respectively, it follows that $p(x + y) \leq p(x) + p(y)$.

Proof of (iii). Suppose $z \notin K$ but $p(z) < 1$. Then there exists an r, $0 \leq r < 1$, such that $z \in rK$. Since $0 \in K$ and K is convex, it follows that $rk = rk + (1 - r)0 \in K$, $k \in K$. Thus $z \in rK \subset K$, which is a contradiction.

Proof of theorem I.5.10. Since $x \notin K$ and K is closed, there exists an r-ball S, $r > 0$, such that $x + S$ and K are disjoint. Thus $k_0 \in K$ implies $x - k_0 \notin -k_0 + K - S = K_1$. Furthermore, $S = -S \subset K_1$, showing that 0 is an interior point of K_1. Since K and S are convex, it follows that K_1 is also convex. Hence the Minkowski functional p of K_1 is defined. We first prove the theorem for the case when X is a real normed linear space. On the one-dimensional space sp $\{x_0\}$, where $x_0 = x - k_0$, define linear functional f by

$$f(\alpha x_0) = \alpha p(x_0)$$

Now, $f(\alpha x_0) \leq p(\alpha x_0)$; for if $\alpha \geq 0$, then $f(\alpha x_0) = p(\alpha x_0)$ by Lemma I.5.12. If $\alpha < 0$, then $f(\alpha x_0) = \alpha p(x_0) \leq 0 \leq p(\alpha x_0)$. Hence, by Remark I.5.3 and Lemma I.5.12, there exists a linear functional F which is an extension of f to all of X such that $F(y) \leq p(y)$, $y \in X$. Moreover, F is bounded; for if $\|y\| = 1$, then $\pm ry \in S \subset K_1$. Hence

$$\pm F(y) = F(\pm y) \leq p(\pm y) \leq r^{-1}$$

Since $x_0 = x - k_0 \notin K_1$ and $-k_0 + k \in -k_0 + K - S = K_1$ for all $k \in K$, it follows from *(iii)* of Lemma I.5.12 and the definition of p that for each $k \in K$,

$$F(x) - F(k_0) = f(x_0) = p(x_0) \geq 1 \geq p(-k_0 + k) \geq -F(k_0) + F(k)$$

Thus

$$F \neq 0 \qquad \text{and} \qquad F(x) \geq F(k), \quad k \in K$$

proving the theorem for real normed linear spaces.

If X is complex, then as in the proof of the Hahn-Banach extension theorem, let X_r be X considered as a real normed linear space. Then by what has just been shown, there exists an $f_r \neq 0$ in X_r' such that $f_r(x) \geq f_r(k)$, $k \in K$. Define f on X by

$$f(x) = f_r(x) - if_r(ix)$$

It is easy to verify that f satisfies the demands of the theorem.

For a treatment of separation theorems for convex sets the reader is referred to Dunford and Schwartz [1], pages 409–418.

I.6 CONJUGATE SPACES

I.6.1 Definition. *Normed linear spaces X and Y are called* ***equivalent*** *if there exists a linear* ***isometry*** *from X* ***onto*** *Y.*

A particular case of the next theorem is essential for the study of differential operators in Chaps. VI and VII. The proof may be found in Dunford and Schwartz [1], pages 286–290.

We say that p' is conjugate to real number p if $1/p + 1/p' = 1$, with the understanding that $p' = \infty$ when $p = 1$.

I.6.2 Theorem. *Let (S, Σ, μ) be a positive measure space. If $1 < p < \infty$, then $\mathfrak{L}'_p(S, \Sigma, \mu)$ is equivalent to $\mathfrak{L}_{p'}(S, \Sigma, \mu)$, p' conjugate to p, in which $x' \in \mathfrak{L}'_p(S, \Sigma, \mu)$ is related to the corresponding $g \in \mathfrak{L}_{p'}(S, \Sigma, \mu)$ by*

$$x'f = \int_S gf\,d\mu \qquad f \in \mathfrak{L}_p(S, \Sigma, \mu)$$

If, in addition, S is the union of a countable number of sets of finite μ-measure, that is, S is σ-finite, then the theorem holds for $p = 1$.

For convenience we write $x' = g$.

As noted in Example I.1.7, $l_p = \mathfrak{L}_p(S, \Sigma, \mu)$, where (S, Σ, μ) is a σ-finite positive measure space. Thus we have the following result.

If $1 \leq p < \infty$, then l'_p is equivalent to $l_{p'}$, p' conjugate to p; $y' \in l'_p$ is related to the corresponding $y = (y_1, y_2, \ldots) \in l_{p'}$ by

$$y'(\alpha) = \sum_{i=1}^{\infty} \alpha_i y_i \qquad \alpha = (\alpha_1, \alpha_2, \ldots) \in l_p$$

For convenience we write $y' = (\alpha_1, \alpha_2, \ldots)$.

I.6.3 Definition. *A set K in normed linear space X is called* **orthogonal** *to a set $F \subset X'$ if $x'k = 0$ for all $k \in K$ and $x' \in F$.*

The reason for the terminology *orthogonal* will be made clear in Sec. I.7.

The **orthogonal complement** *in X' of K, denoted by $K^\perp$, is the set of elements in X' which is orthogonal to K.*

Even if K is not a subspace, $K^\perp$ is a closed subspace of X'.

I.6.4 Theorem. *Let M be a subspace of normed linear space X. Then*

i. $X'/M^\perp$ is equivalent to M' under the map U defined by $U[x'] = x'_M$ where $[x']$ is in $X'/M^\perp$ and x'_M is the restriction of x' to M.

ii. If M is closed (so that X/M is a normed linear space), then $(X/M)'$ is equivalent to $M^\perp$ under the map V defined by

$$(Vz')x = z'[x] \qquad z' \in \left(\frac{X}{M}\right)'$$

Proof of (i). Note that U is unambiguously defined, since $[y'] = [x']$ implies $0 = y'm - x'm$, $m \in M$. Clearly, U is linear with range in M'. Given $m' \in M'$, there exists, by Corollary I.5.4, an $x' \in X'$ which is an extension of m'. Hence $U[x'] = x'_M = m'$ which shows that $\mathcal{R}(U) = M'$. For any $y' \in [x']$,

$$\|U[x']\| = \|y'_M\| \leq \|y'\|$$

Thus

$$\|U[x']\| \leq \inf_{y' \in [x']} \|y'\| = \|[x']\| \tag{1}$$

On the other hand, there exists a $v' \in X'$ which is an extension of x'_M such that $\|v'\| = \|x'_M\|$. Therefore v' is in $[x']$ and

$$\|U[x']\| = \|x'_M\| = \|v'\| \geq \|[x']\| \tag{2}$$

By (1) and (2), $\|U[x']\| = \|[x']\|$.

Proof of (ii). For $z' \in (X/M)'$,

$$|(Vz')x| = |z'[x]| \leq \|z'\| \, \|[x]\| \leq \|z'\| \, \|x\| \qquad x \in X$$

and

$$(Vz')m = z'[m] = z'[0] = 0 \qquad m \in M$$

Thus Vz' is in $M^\perp$ with

$$\|Vz'\| \leq \|z'\| \tag{3}$$

Since

$$|z'[x]| = |(Vz')y| \leq \|Vz'\| \, \|y\| \qquad y \in [x]$$

it follows that

$$|z'[x]| \leq \|Vz'\| \, \|[x]\|$$

Thus

$$\|z'\| \leq \|Vz'\|$$

which, together with (3), proves that V is an isometry. Given $x' \in M^\perp$, let z' be the linear functional on X/M defined by $z'[x] = x'x$. Since

$$|z'[x]| = |x'y| \leq \|x'\| \, \|y\| \qquad y \in [x]$$

it follows that $|z'[x]| \leq \|x'\| \, \|[x]\|$. Hence z' is in $(X/M)'$. Furthermore, $Vz' = x'$, proving that $\mathcal{R}(V) = M^\perp$.

I.6.5 Definition. *A set A in a normed linear space is called* **separable** *if there exists a countable subset of A which is dense in A.*

I.6.6 Example. l_p, $1 \leq p < \infty$, is separable. A countable subset which is dense in l_p is the set of all sequences of the form $(\alpha_1, \alpha_2, \ldots,$

$\alpha_k, 0, 0, \ldots)$, where the α_k are rational (a complex number is rational if its real and imaginary parts are rational).

l_∞ is not separable. For suppose $\{x_k\}$ is a sequence of elements in l_∞. We show that there exists an $x \in l_\infty$ such that $\|x - x_k\| \geq 1$. Let

$$\begin{aligned} x_1 &= (x_1{}^1, x_2{}^1, \ldots) \\ x_2 &= (x_1{}^2, x_2{}^2, \ldots) \\ &\cdots\cdots\cdots\cdots \\ x_k &= (x_1{}^k, x_2{}^k, \ldots) \end{aligned}$$

Define $x = (\alpha_1, \alpha_2, \ldots)$ by

$$\alpha_k = \begin{cases} 0 & \text{if } |x_k{}^k| \geq 1 \\ 1 + |x_k{}^k| & \text{if } |x_k{}^k| < 1 \end{cases}$$

Then x is in l_∞ and $\|x - x_j\| \geq |\alpha_j - x_j{}^j| \geq 1$.

It is shown in Zaanen [1], page 75, that $\mathfrak{L}_p(S, \Sigma, \mu)$, $1 \leq p < \infty$, is separable, where μ is a separable measure. In particular, $\mathfrak{L}_p(\Omega)$, $1 \leq p < \infty$, is separable, where Ω is a Lebesgue-measurable subset of E^n. Taylor [1], page 91, shows that $\mathfrak{L}_\infty(I)$ is not separable, where I is any interval on the line.

I.6.7 Theorem. *If the conjugate of normed linear space X is separable, then X is also separable.*

Proof. Let $\{x_k'\}$ be dense in X'. Choose sequence $\{x_k\}$ in X such that $\|x_k\| = 1$ and $|x_k'x_k| \geq \|x_k'\|/2$. We claim that the space X_1 spanned by the x_k is dense in X. Suppose this is not the case. Then, by Corollary I.5.5, there exists an $x' \neq 0$ in X' such that $x'X_1 = 0$. Since $\{x_k'\}$ is dense in X', there exists a subsequence $\{x_{k_i}'\}$ which converges to x'. Since

$$\|x_{k_i}' - x'\| \geq |(x_{k_i}' - x')x_{k_i}| = |x_{k_i}'x_{k_i}| \geq \frac{\|x_{k_i}'\|}{2}$$

$\{x_{k_i}'\}$ converges to 0. But this is impossible since $x_{k_i}' \to x' \neq 0$. Thus X_1 is dense in X. Since the set A, consisting of elements of the form $\sum_{i=1}^{n} \alpha_i x_i$, α_i rational, is countable and dense in X_1, it follows that A is dense in X. Hence X is separable.

Since $l_\infty = l_1'$ is not separable but l_1 is, the converse to the above theorem does not hold.

I.6.8 Definition. *The **natural map,** denoted by J_X, of normed linear space X into its second conjugate space X'' (the Banach space of bounded* *linear functionals on X') is defined by*

$$(J_X x)x' = x'x \qquad x' \epsilon X'$$

If the range of J_X is all of X'', then X is called ***reflexive.***

I.6.9 Remarks

i. The natural map J from X into X'' is a linear isometry. The linearity of J is clear, while from Corollary I.5.7 we have

$$\|Jx\| = \sup_{\|x'\|=1} |(Jx)x'| = \sup_{\|x'\|=1} |x'x| = \|x\|$$

ii. Every reflexive space is complete, since a conjugate space is complete and isomorphisms preserve completeness.

iii. $\mathfrak{L}_p(S, \Sigma, \mu)$, $1 < p < \infty$, is reflexive, as may easily be seen from the definition of the natural map and the representation of $\mathfrak{L}'_p(S, \Sigma, \mu)$ given in Theorem I.6.2.

A word of caution. Sometimes the argument given to show, for example, that $\mathfrak{L}_p = \mathfrak{L}_p(S, \Sigma, \mu)$, $1 < p < \infty$, is reflexive is the following. $\mathfrak{L}''_p$ is equivalent to $\mathfrak{L}'_{p'}$, which in turn is equivalent to $\mathfrak{L}_{p''} = \mathfrak{L}_p$. Hence $\mathfrak{L}_p$ is reflexive.

The flaw in the argument is that the equivalence of a normed linear space with its second conjugate does not guarantee the reflexivity of the space. James [1] gave an ingenious construction of a Banach space X which is equivalent to X'', yet the dimension of $X''/J_X X$ is 1.

I.6.10 Theorem. *The conjugate space of a separable reflexive space is separable.*

Proof. Suppose X is reflexive and separable. Then $X'' = J_X X$ is separable. Hence X' is separable by Theorem I.6.7.

Since $l'_1 = l_\infty$ and l_1 is separable but l_∞ is not, it follows from the above theorem that l_1 is not reflexive.

I.6.11 Theorem. *A Banach space is reflexive if and only if its conjugate is reflexive.*

Proof. Let J be the natural map from X into X''. Suppose X is reflexive. We must show that given $x''' \epsilon X'''$, there exists an $x' \epsilon X'$ such that

$$x'''Jx = (Jx)x' = x'x \qquad x \epsilon X \tag{1}$$

Define $x' = x'''J$. Then $x' \epsilon X'$ and (1) is satisfied.

Assume X' is reflexive but X is not reflexive. There exists an $x'' \in X''$ such that $x'' \notin JX$. Since X is complete and J is a linear isometry, we know from Lemma I.4.8 that JX is also complete and therefore is closed in X''. Thus there exists an $x''' \in X'''$ such that $x'''x'' \neq 0$ and $x'''JX = 0$. Since X' is reflexive, there exists an $x' \in X'$ such that

$$(2) \qquad 0 \neq x'''x'' = x''x'$$

and

$$(3) \qquad 0 = x'''Jx = (Jx)x' = x'x \qquad x \in X$$

Equation (2) shows that $x' \neq 0$, while Eq. (3) shows that $x' = 0$, which is absurd. Hence X is reflexive.

I.6.12 Theorem. *A closed subspace of a reflexive space is reflexive.*

Proof. Let M be a closed subspace of reflexive space X. Given $m'' \in M''$, define $x'' \in X''$ by

$$x''x' = m''x'_M$$

where x'_M is the restriction of $x' \in X'$ to M. Let $m = J_X^{-1}x''$. It will be shown that m is in M and $J_M m = m''$, proving that M is reflexive. Suppose $m \notin M$. Then there exists an $x' \in X'$ such that $x'm \neq 0$ while $x'M = 0$. Thus $x'_M = 0$ and

$$0 \neq x'm = x'J_X^{-1}x'' = x''x' = m''x'_M = m''0 = 0$$

which is impossible. Hence m is in M. For each $m' \in M'$, let m'_e be an element in X', which is an extension of m'. Then

$$m''m' = x''m'_e = m'_e J_X^{-1}x'' = m'_e m = m'm$$

Thus $J_M m = m''$, completing the proof of the theorem.

The next theorem is useful in proving convergence in norm of certain sequences in $\mathfrak{L}_p(\Omega)$, $1 < p < \infty$.

I.6.13 Definition. *A sequence $\{x_n\}$ in normed linear space X is said to converge **weakly** to $x \in X$ if for every $x' \in X'$, $x'x_n \to x'x$. This is written $x_n \xrightarrow{w} x$.*

I.6.14 Remark. Banach [1], pages 137–139, has shown that in l_1, convergence in norm is the same as weak convergence.

I.6.15 Theorem. *Every bounded sequence in a reflexive space contains a weakly convergent subsequence.*

Proof. Let $\{x_k\}$ be a bounded sequence in reflexive space X and let $X_1 = \overline{\text{sp}}\ \{x_k\}$. Then X_1 is separable, since each $x \in X_1$ is the limit of elements of the form $\sum_{i=1}^{n} \alpha_i x_i$, α_i rational. Thus X_1' is separable by Theorems I.6.12 and I.6.10. Let $\{x_k'\}$ be dense in X_1'. Since $\{x_1' x_k\}$ is a bounded sequence, there exists a subsequence $\{x_1' x_{1k}\}$ which converges. Since $\{x_2' x_{1k}\}$ is bounded, there exists a subsequence $\{x_2' x_{2k}\}$ which converges. Thus $\{x_1' x_{2k}\}$ and $\{x_2' x_{2k}\}$ converge. By induction, one obtains sequences $\{x_{jk}\}$ such that

(1) $$\{x_{(j+1)k}\} \text{ is a subsequence of } \{x_{jk}\} \qquad 1 \leq j$$

(2) $$\{x_j' x_{jk}\} \text{ converges}$$

Setting $v_k = x_{kk}$, it is easy to see from (1) and (2) that $\{x_j' v_k\}$ converges for each j. Since $\{x_j'\}$ is dense in X_1', it follows that $\{x' v_k\}$ converges for each $x' \in X_1'$. Define x'' on X_1' by

$$x'' x' = \lim_{k \to \infty} x' v_k$$

Clearly, x'' is linear. Since $|x' v_k| \leq \|x'\|\ \|v_k\|$ and $\{v_k\}$ is a subsequence of bounded sequence $\{x_k\}$, it follows that x'' is bounded; that is, $x'' \in X_1''$. Therefore the reflexivity of X_1 implies the existence of an $x \in X_1$ such that

(3) $$x' x = x'' x' = \lim_{k \to \infty} x' v_k \qquad x' \in X_1'$$

Given $z' \in X'$, let z_R' be the restriction of z' to X_1. By (3),

$$z' x = z_R' x = \lim_{k \to \infty} z' v_k$$

Hence $\{v_k\}$ converges weakly to x, which proves the theorem.

Convergence in norm is not the same as weak convergence in an infinite-dimensional reflexive space X. Otherwise, by the above theorem, every sequence in the 1-sphere of X contains a convergent (in norm) subsequence. Thus the 1-sphere is compact, and therefore X is finite-dimensional.

The properties of the natural map from X into X'' lead to a simple proof of the following theorem.

I.6.16 Theorem. *Every normed linear space is a dense subspace of a Banach space.*

Proof. Given normed linear space X, let $\hat{X}$ consist of the elements of X and the elements of the Banach space $Z = \overline{J_X X} \subset X''$ which are not in $J_X X$. Define addition and scalar multiplication in $\hat{X}$ by the corresponding operations in Z. The norm of an element in $\hat{X}$ is the norm of the corresponding element in Z.

We shall call $\hat{X}$ the completion of X.

I.7 HILBERT SPACES

In this section some of the basic theorems concerning Hilbert spaces are proved. The study of these spaces has been the subject of several books. The reader is referred to Akhieser and Glazman [1], Berberian [1], Halmos [1], and Stone [1].

I.7.1 Definition. *Let X be a vector space over the real or complex numbers. An inner product on X is a scalar-valued function $\langle\ ,\ \rangle$ defined on the Cartesian product $X \times X$ with the following properties.*

i. $\langle \alpha x, y\rangle = \alpha\langle x, y\rangle$
ii. $\langle x, y\rangle = \overline{\langle y, x\rangle}$; that is, $\langle x, y\rangle$ is the complex conjugate of $\langle y, x\rangle$
iii. $\langle x + y, z\rangle = \langle x, z\rangle + \langle y, z\rangle$
iv. $\langle x, x\rangle > 0$ *whenever* $x \neq 0$

X, together with an inner product, is called an ***inner-product space.***

It follows from the above axioms that

$$\langle x, \alpha y\rangle = \bar{\alpha}\langle x, y\rangle$$

$$\langle x, y + z\rangle = \langle x, y\rangle + \langle x, z\rangle$$

$$\langle 0, x\rangle = 0 \qquad x \in X$$

I.7.2 Example. For unitary n-space an inner product is given by

$$\langle x, y\rangle = \sum_{i=1}^{n} x_i\bar{y}_i \qquad x = (x_1, \ldots, x_n), \quad y = (y_1, \ldots, y_n)$$

I.7.3 Example. An inner product on $\mathfrak{L}_2(\Omega)$ is given by

$$\langle f, g\rangle = \int_\Omega f\bar{g}\, d\mu$$

In Examples I.7.2 and I.7.3, $\langle x, x\rangle^{\frac{1}{2}}$ gives the norm of x. We shall show that for any inner-product space, $\|x\| = \langle x, x\rangle^{\frac{1}{2}}$ is indeed a norm.

I.7.4 Schwarz's inequality. *Let X be an inner-product space. Then for all x and $y \in X$*

$$|\langle x, y\rangle| \leq \|x\| \, \|y\| \qquad \|x\| = \langle x, x\rangle^{\frac{1}{2}}$$

Equality holds if and only if x and y are linearly dependent.

Two proofs are given. The first one is the usual short proof, while the second, though longer, is geometrically motivated.

Proof. For any scalar λ,

$$(1) \qquad 0 \leq \langle x - \lambda y, x - \lambda y\rangle = \|x\|^2 - \bar{\lambda}\langle x, y\rangle - \lambda\langle y, x\rangle + |\lambda|^2 \, \|y\|^2$$

For $\langle y, x\rangle = 0$, the inequality is trivial and equality holds if and only if at least x or y is zero. Suppose $\langle y, x\rangle \neq 0$. Substituting $\lambda = \|x\|^2/\langle y, x\rangle$ in (1) yields

$$(2) \qquad 0 \leq -\|x\|^2 + \frac{\|x\|^4 \, \|y\|^2}{|\langle x, y\rangle|^2}$$

whence

$$(3) \qquad |\langle x, y\rangle|^2 \leq \|x\|^2 \, \|y\|^2$$

Equality holds in (3) if and only if equality holds in (1), in which case $x = (\|x\|^2/\langle y, x\rangle)y$.

In the plane, Schwarz's inequality follows from the fact that $\langle x, y\rangle = \|x\| \, \|y\| \cos \theta$, where θ is the angle between x and y, $0 \leq \theta \leq \pi$. The first step of the proof is to give meaning to $\cos \theta$ for an arbitrary inner-product space. To avoid trivialities, it is assumed in the sequel that $x \neq 0$ and $y \neq 0$. Choose λ so that $x - \lambda y$ is perpendicular to y; that is, $0 = \langle x - \lambda y, y\rangle$, or equivalently,

$$\lambda = \frac{\langle x, y\rangle}{\|y\|^2}$$

Thus we may think of a right triangle with hypotenuse x and sides λy and $x - \lambda y$. Define

$$(1) \qquad \begin{aligned} \cos \theta &= \frac{\|\lambda y\|}{\|x\|} \\ \sin \theta &= \frac{\|x - \lambda y\|}{\|x\|} \end{aligned}$$

(Envision θ as being the acute angle between x and y.) A simple computation shows that

$$(2) \qquad |\langle x, y\rangle| = \|x\| \, \|y\| \cos \theta$$

From

$$\|x\|^2 = \langle x - \lambda y + \lambda y, x - \lambda y + \lambda y\rangle$$

and the fact that λ was chosen so that $\langle x - \lambda y, y\rangle = 0$, it follows that the "Pythagorean theorem" holds; i.e.,

$$(3) \qquad \|x\|^2 = \|x - \lambda y\|^2 + \|\lambda y\|^2$$

Thus (1) and (3) imply

$$(4) \qquad \cos^2 \theta + \sin^2 \theta = 1$$

Hence $0 \leq \cos \theta \leq 1$ which, together with (2), implies

$$|\langle x, y\rangle| \leq \|x\| \, \|y\|$$

Suppose $|\langle x, y\rangle| = \|x\| \, \|y\|$. Then (2) and (4) show that

$$0 = \sin \theta = \frac{\|x - \lambda y\|}{\|x\|}$$

or $x = \lambda y$.

I.7.5 Theorem. *Given inner-product space X, $\|x\| = \langle x, x\rangle^{\frac{1}{2}}$ defines a norm on X.*

Proof. For x and $y \in X$, Schwarz's inequality implies

$$\begin{aligned}\|x + y\|^2 &= \langle x + y, x + y\rangle = \|x\|^2 + \langle x, y\rangle + \overline{\langle x, y\rangle} + \|y\|^2 \\ &\leq \|x\|^2 + 2|\langle x, y\rangle| + \|y\|^2 \leq \|x\|^2 + 2\|x\| \, \|y\| + \|y\|^2 \\ &= (\|x\| + \|y\|)^2\end{aligned}$$

Hence $\|x + y\| \leq \|x\| + \|y\|$. The other properties required of $\| \quad \|$ in order that it be a norm follow immediately from the definition of an inner product.

An inner-product space will always be considered as a normed linear space with the norm $\|x\| = \langle x, x\rangle^{\frac{1}{2}}$.

I.7.6 Theorem. *Suppose $x_n \to x$ and $y_n \to y$ in inner-product space X. Then $\langle x_n, y_n\rangle \to \langle x, y\rangle$. In other words, the inner product is continuous on the metric space $X \times X$.*

Proof. We write

$$\langle x_n, y_n \rangle - \langle x, y \rangle = \langle x_n, y_n \rangle - \langle x, y_n \rangle + \langle x, y_n \rangle - \langle x, y \rangle$$
$$= \langle x_n - x, y_n \rangle + \langle x, y_n - y \rangle$$

By Schwarz's inequality

$$|\langle x_n, y_n \rangle - \langle x, y \rangle| \le \|x_n - x\| \, \|y_n\| + \|x\| \, \|y_n - y\|$$

Since $\|y_n\| \to \|y\|$, $\|x_n - x\| \to 0$ and $\|y_n - y\| \to 0$, it follows that $\langle x_n, y_n \rangle - \langle x, y \rangle \to 0$.

I.7.7 Definition. *A* ***Hilbert space*** *is an inner-product space which is also a Banach space with norm* $\|x\| = \langle x, x \rangle^{\frac{1}{2}}$. *Thus* $\mathfrak{L}_2(S, \Sigma, \mu)$ *is a Hilbert space, but any nonclosed subspace of* $\mathfrak{L}_2(S, \Sigma, \mu)$ *is an inner-product space which is not a Hilbert space.*

In the course of solving the following problem, the theorems which we desire to include in this section make their appearance.

I.7.8 Problem. Suppose M is a closed subspace of Hilbert space $\mathfrak{H}$. Given $x \in \mathfrak{H}$, does there exist an element $Px \in M$ such that

$$\|x - Px\| = d(x, M)$$

In other words, does there exist a best approximation to x among all the elements in M—best in the sense that $\|x - Px\| \le \|x - z\|$ for all $z \in M$?

M must be closed, for if $x \in \bar{M}$ but $x \notin M$, then for all $m \in M$,

$$d(x, M) = 0 < \|x - m\|$$

If M is a line through the origin and $\mathfrak{H}$ is the plane, then Px is the foot of the perpendicular from x to M or, what is the same thing, $x - Px$ is perpendicular to M. The next lemma is motivated by this observation.

I.7.9 Definition. *Elements x and y in an inner-product space are called* ***perpendicular*** *or* ***orthogonal,*** *written* $x \perp y$, *when* $\langle x, y \rangle = 0$. *Element x is perpendicular to a set M, written* $x \perp M$, *if* $\langle x, m \rangle = 0$ *for all* $m \in M$

I.7.10 Lemma. *Let X be an inner-product space (not necessarily complete) and let M be a subspace of X. Suppose that given $x \in X$, there exists a $Px \in M$ such that $x - Px \perp M$. Then Px is the unique element in M for which $d(x, M) = \|x - Px\|$. More specifically, the "Pythagorean theorem" holds; that is,*

$$\|x - m\|^2 = \|x - Px\|^2 + \|Px - m\|^2 \qquad m \in M$$

Proof. It is clear that the Pythagorean theorem implies that

$$\|x - m\| > \|x - Px\| = d(x, M) \qquad m \neq Px, \quad m \in M$$

Given $m \in M$,

$$\begin{aligned} (1) \quad \|x - m\|^2 &= \langle x - m, x - m\rangle \\ &= \langle (x - Px) + (Px - m), (x - Px) + (Px - m)\rangle \\ &= \|x - Px\|^2 + \langle x - Px, Px - m\rangle \\ &\qquad + \langle Px - m, x - Px\rangle + \|Px - m\|^2 \end{aligned}$$

Since $x - Px \perp M$ and $Px - m \in M$, it follows from (1) that

$$\|x - m\|^2 = \|x - Px\|^2 + \|Px - m\|^2$$

Problem I.7.8 is now reduced to finding the "foot of the perpendicular" Px. We start out by finding Px for certain M.

I.7.11 Definition. *A set in an inner-product space is called* **orthogonal** *if distinct elements in the set are orthogonal.*

An orthogonal set is called **orthonormal** *if each element in the set has norm* 1.

In l_2 the set of unit vectors $u_k = (0, 0, \ldots, \underset{k}{1}, 0, \ldots)$ is orthonormal. For the $\mathcal{L}_2([0, 2\pi])$ space of complex-valued functions, $\{(1/\sqrt{2\pi})e^{int}\}$, $n = 0, \pm 1, \pm 2, \ldots$, is an orthonormal set.

In order to motivate how to determine Px in the next lemma, take $\mathcal{H} = E^3$ with M the plane spanned by the vectors (1, 0, 0) and (0, 1, 0). For $x = (x_1, x_2, x_3)$ the foot of the perpendicular is

$$Px = (x_1, x_2, 0) = x_1u_1 + x_2u_2 = \langle x, u_1\rangle u_1 + \langle x, u_2\rangle u_2$$

Also, the length of Px is dominated by the length of x (Bessel's inequality).

I.7.12 Lemma. *Let $\{u_1, u_2, \ldots\}$ be an orthonormal set in Hilbert space $\mathcal{H}$ and let $M = \overline{\text{sp}}\,(\{u_k\})$. Then for each $x \in \mathcal{H}$ there exists a Px in M such that $x - Px \perp M$. The formula for Px is given by*

$$Px = \sum_{i=1}^{\infty} \langle x, u_i\rangle u_i$$

The series converges to Px independent of rearrangement. Furthermore,

$$\sum_{i=1}^{\infty} |\langle x, u_i\rangle|^2 = \|Px\|^2 \leq \|x\|^2 \qquad \textit{(Bessel's inequality)}$$

Proof. The convergence of the series will be established first. Since $\{u_1, u_2, \ldots\}$ is orthonormal, it follows that

$$0 \leq \langle x - \sum_{i=1}^{n} \langle x, u_i \rangle u_i, x - \sum_{i=1}^{n} \langle x, u_i \rangle u_i \rangle = \|x\|^2 - \sum_{i=1}^{n} |\langle x, u_i \rangle|^2$$

Hence

$$\sum_{i=1}^{\infty} |\langle x, u_i \rangle|^2 \leq \|x\|^2 \tag{1}$$

Thus for $s_n = \sum_{i=1}^{n} \langle x, u_i \rangle u_i$

$$\|s_{n+k} - s_n\|^2 = \langle \sum_{i=n+1}^{n+k} \langle x, u_i \rangle u_i, \sum_{i=n+1}^{n+k} \langle x, u_i \rangle u_i \rangle$$

$$= \sum_{i=n+1}^{n+k} |\langle x, u_i \rangle|^2 \to 0 \quad \text{as } n \to \infty$$

Therefore $\{s_n\}$ converges to an element Px in $\mathcal{H}$. Note that the argument also shows that the series converges independent of any rearrangement. Since s_n is in M and M is closed, Px is also in M. Now, by Theorem I.7.6,

$$\langle x - Px, u_j \rangle = \lim_{n \to \infty} \langle x - \sum_{i=1}^{n} \langle x, u_i \rangle u_i, u_j \rangle = \langle x, u_j \rangle - \langle x, u_j \rangle = 0 \qquad 1 \leq j$$

Hence $x - Px \perp \text{sp}\,(\{u_i\})$ and therefore, by the continuity of the inner product, $x - Px \perp M$. By (1),

$$\|Px\|^2 = \lim_{n \to \infty} \langle \sum_{i=1}^{n} \langle x, u_i \rangle u_i, \sum_{i=1}^{n} \langle x, u_i \rangle u_i \rangle = \sum_{i=1}^{\infty} |\langle x, u_i \rangle|^2 \leq \|x\|^2$$

Suppose that $P_1 x = \sum_{i=1}^{\infty} \langle x, v_i \rangle v_i$, where $\{v_i\}$ is a rearrangement of $\{u_i\}$. Then by the above argument, $x - P_1 x \perp M$. Hence, by Lemma I.7.10,

$$\|x - Px\| = d(x, M) = \|x - P_1 x\| \qquad \text{and} \qquad P_1 x = Px$$

Problem I.7.8 is therefore solved for $M = \overline{\text{sp}}\,\{u_k\}$. The next step is to solve the problem for $M = \overline{\text{sp}}\,\{u_\alpha\}$, where $\{u_\alpha\}$ is not necessarily countable.

I.7.13 Lemma. *Let $M = \overline{\text{sp}}\ \{u_\alpha\}$, where $\{u_\alpha\}$ is an orthonormal set in Hilbert space $\mathcal{H}$. Then for each $x \in \mathcal{H}$, $\langle x, u_\alpha \rangle \neq 0$ for at most a countable number of u_α, and*

$$Px = \sum_{\langle x, u_\alpha \rangle \neq 0} \langle x, u_\alpha \rangle u_\alpha$$

The series converges to $Px \in M$ independent of rearrangement. Moreover, $x - Px \perp M$.

Proof. Once it is shown that $\langle x, u_\alpha \rangle \neq 0$ for at most a countable number of u_α, the rest of the lemma follows from Lemma I.7.12. For each sequence of elements $\{u_{\alpha_i}\}$ in $\{u_\alpha\}$, we have, by Bessel's inequality, that $\sum_{i=1}^{\infty} |\langle x, u_{\alpha_i} \rangle|^2 \leq \|x\|^2$. Consequently, there exist, for each positive integer n, at most n of the $\langle x, u_\alpha \rangle$ such that $|\langle x, u_\alpha \rangle| \geq \|x\|^2/n$. Define

$$E_n = \left\{ u_\alpha \;\middle|\; |\langle x, u_\alpha \rangle|^2 \geq \frac{\|x\|^2}{n} \right\}$$

Then

$$\left\{ u_\alpha \;\middle|\; |\langle x, u_\alpha \rangle| > 0 \right\} = \bigcup_{n=1}^{\infty} E_n$$

is countable, since each E_n is finite.

The class of those M for which there exists a solution to Problem I.7.8 has now been enlarged to those $M = \overline{\text{sp}}\ \{u_\alpha\}$, where $\{u_\alpha\}$ is an arbitrary orthonormal set. Thus, to solve the problem for general closed subspaces M, it remains to show that M is the closure of the span of some orthonormal set.

I.7.14 Lemma. *Given any closed supspace $M \neq 0$ of Hilbert space $\mathcal{H}$, there exists an orthonormal set $\{u_\alpha\}$ such that $M = \overline{\text{sp}}\ \{u_\alpha\}$.*

Proof. The proof utilizes Zorn's lemma in order to prove the existence of a maximal orthonormal set O contained in M, maximal in the sense that O is not a proper subset of any orthonormal set contained in M. Let $\mathcal{P}$ be the class of all orthonormal subsets of M. $\mathcal{P}$ is not empty since it contains the 1-point set $\{m\}$, where $m \in M$ and $\|m\| = 1$. Partially order $\mathcal{P}$ by set inclusion. Let $\mathfrak{I}$ be a totally ordered subclass of $\mathcal{P}$. Since $\bigcup_{S \in \mathfrak{I}} S$ is an upper bound in $\mathcal{P}$ for $\mathfrak{I}$, Zorn's lemma asserts that $\mathcal{P}$ contains a maximal element, say the orthonormal set $\{u_\alpha\}$. We claim that $M = \overline{\text{sp}}\ \{u_\alpha\}$. Assume the contrary. Then there exists an $x \neq 0$ in M but not in $M_1 = \overline{\text{sp}}\ \{u_\alpha\}$. By Lemma I.7.13, there exists a $Px \in M_1$ such that $x - Px \perp M_1$. Thus the set consisting of $(x - Px)/\|x - Px\|$ and the u_α is an orthonormal set containing $\{u_\alpha\}$ properly. But this is impossible, since $\{u_\alpha\}$ is maximal. Hence $M = \overline{\text{sp}}\ \{u_\alpha\}$.

From the preceding lemmas we have the following theorem.

I.7.15 Theorem. *Let M be a closed subspace of Hilbert space $\mathcal{H}$. Given $x \in \mathcal{H}$, there exists a unique element $Px \in M$ such that $x - Px \perp M$. Furthermore, $\|x - Px\| = d(x, M) < \|x - m\|$, $m \in M$, $m \neq Px$.*

I.7.16 Definition. *Subspaces M_1 and M_2 of vector space V are called* ***linearly independent*** *if $M_1 \cap M_2 = (0)$ or, equivalently, each $x \in M_1 + M_2$ has a unique representation $x = x_1 + x_2$, $x_1 \in M_1$, $x_2 \in M_2$.*

The symbol $M_1 \oplus M_2$, called the direct sum of M_1 and M_2, is used only if M_1 and M_2 are linearly independent and denotes the set $M_1 + M_2$.

I.7.17 Corollary. *If M is a closed subspace of Hilbert space $\mathcal{H}$, then*

$$\mathcal{H} = M \oplus M^{\perp}$$

where $M^{\perp}$ is the closed subspace of elements perpendicular to M.

Proof. If $y \in M \cap M^{\perp}$, then $0 = \langle y, y \rangle$ and therefore $y = 0$. Given $x \in \mathcal{H}$, there exists, by Theorem I.7.15, a $Px \in M$ such that $x - Px \perp M$. Since

$$x = Px + (x - Px) \in M + M^{\perp}$$

the corollary is proved.

For a fixed x in Hilbert space $\mathcal{H}$ the functional x' defined by $x'(z) = \langle z, x \rangle$ is in $\mathcal{H}'$. The question arises whether all elements in $\mathcal{H}'$ are determined in this manner. The answer is in the affirmative, as is now shown.

I.7.18 Riesz representation theorem. *Let $\mathcal{H}$ be a Hilbert space. Given $x' \in \mathcal{H}'$, there exists a unique $x \in \mathcal{H}$ such that*

$$x'(z) = \langle z, x \rangle \qquad z \in \mathcal{H}$$

Moreover, $\|x\| = \|x'\|$.

Proof. If $x' = 0$, choose $x = 0$. Assume $x' \neq 0$. If x were to exist as in the theorem, then

$$0 = x'(z) = \langle x, z \rangle \qquad z \in \mathfrak{N}(x') \tag{1}$$

The first step then is to find an element $v \neq 0$ orthogonal to $\mathfrak{N}(x')$. Since $\mathfrak{N}(x')$ is closed and is not all of $\mathcal{H}$, such a v exists by Theorem I.7.15. For any scalar α, $x = \alpha v$ is also orthogonal to $\mathfrak{N}(x')$. We need another condition to determine which α to choose. This condition is given by

$$(2) \qquad x'(x) = \langle x, x \rangle$$

or

$$\alpha x'(v) = \langle \alpha v, \alpha v \rangle = \alpha\bar{\alpha}\|v\|^2$$

Hence, choose $\alpha = \overline{x'(v)}/\|v\|^2$. We assert that $x = \alpha v = (\overline{x'(v)}/\|v\|^2)v$ is the desired x. Since the map $[u] \to x'u$ is a 1-1 linear map from $\mathcal{H}/\mathfrak{N}(x')$ onto the scalars, $\mathcal{H}/\mathfrak{N}(x')$ is one-dimensional and therefore $[x] \neq [0]$ is a basis. Hence $\mathcal{H} = \text{sp}\,\{x\} \oplus \mathfrak{N}(x')$. Given $z \in \mathcal{H}$, z has the representation $z = \alpha x + w$, for some $w \in \mathfrak{N}(x')$. By the choice of x, (1) and (2) are satisfied. Thus

$$x'(z) = \alpha x'(x) = \alpha\langle x, x \rangle = \langle \alpha x + w, x \rangle = \langle z, x \rangle$$

Suppose $x'(z) = \langle z, y \rangle$ for all $z \in \mathcal{H}$. Then $0 = \langle z, x - y \rangle$, $z \in \mathcal{H}$, and, in particular, $0 = \langle x - y, x - y \rangle$. Hence $x = y$. It remains to prove that $\|x\| = \|x'\|$. Since

$$|x'(z)| = |\langle z, x \rangle| \leq \|z\|\,\|x\| \qquad z \in \mathcal{H}$$

and

$$\|x'\|\,\|x\| \geq x'x = \langle x, x \rangle = \|x\|^2$$

it follows that $\|x\| = \|x'\|$.

I.7.19 Definition. *For $\mathcal{H}$ a Hilbert space, let $E_{\mathcal{H}}$ be the map from $\mathcal{H}$ onto $\mathcal{H}'$ defined by $E_{\mathcal{H}}x = x'$, where $x'z = \langle z, x \rangle$, $z \in \mathcal{H}$.*

It is clear that $E_{\mathcal{H}}$ is additive and that $E_{\mathcal{H}}(\alpha x) = \bar{\alpha}E_{\mathcal{H}}x$. By the Riesz representation theorem we know that $E_{\mathcal{H}}$ is an isometry with $\mathfrak{R}(E_{\mathcal{H}}) = \mathcal{H}'$. Thus if $\mathcal{H}$ is a real Hilbert space, then $\mathcal{H}$ is equivalent to $\mathcal{H}'$.

I.7.20 Definition. *A Banach space X with norm $\|\ \|_1$ is called a Hilbert space if there exists an inner product on X such that $\langle x, x \rangle^{\frac{1}{2}} = \|x\|_1$.*

I.7.21 Theorem. *If $\mathcal{H}$ is a Hilbert space, then so is $\mathcal{H}'$. Furthermore, $\mathcal{H}$ is reflexive.*

Proof. Let $E = E_{\mathcal{H}}$ be the map from $\mathcal{H}$ onto $\mathcal{H}'$, as in Definition I.7.19. Define an inner product $\langle\ ,\ \rangle_1$ on $\mathcal{H}'$ by

$$\langle x', y' \rangle_1 = \langle E^{-1}y', E^{-1}x' \rangle$$

where $\langle\ ,\ \rangle$ is the inner product on $\mathcal{H}$. From the properties of E, it is easy to verify that $\langle\ ,\ \rangle_1$ is indeed an inner product. Since E^{-1} is an isometry,

$$\langle x', x' \rangle_1 = \langle E^{-1}x', E^{-1}x' \rangle = \|E^{-1}x'\|^2 = \|x'\|^2$$

Thus $\mathcal{H}'$ is a Hilbert space. Let $E_1 = E_{\mathcal{H}'}$ be the map from $\mathcal{H}'$ onto $\mathcal{H}''$, as in Definition I.7.19. Suppose $x'' \in \mathcal{H}''$. For each $x' \in \mathcal{H}'$,

$$x''x' = \langle x', E_1^{-1}x'' \rangle_1 = \langle E^{-1}E_1^{-1}x'', E^{-1}x' \rangle = x'(E^{-1}E_1^{-1}x'')$$

Thus $\mathcal{H}$ is reflexive.

We conclude this section with the proof that all separable Hilbert spaces are equivalent to l_2.

I.7.22 Lemma. *Let $\mathcal{O}$ be an orthonormal set in Hilbert Space $\mathcal{H}$. The following statements are equivalent.*

i. $\mathcal{O}$ *is a maximal orthonormal set in* $\mathcal{H}$
ii. $\overline{\text{sp}}\,(\mathcal{O}) = \mathcal{H}$
iii. $x = Px = \sum_{u \in \mathcal{O}} \langle x, u \rangle u$
iv. $\|x\| = \|Px\|$ *(Parseval's equality)*

Proof. (*i*) implies (*ii*). If $\overline{\text{sp}}\,(\mathcal{O}) \neq \mathcal{H}$, then it follows from Theorem I.7.15 that there exists a $y \in \mathcal{H}$ such that $\|y\| = 1$, $y \perp \overline{\text{sp}}\,(\mathcal{O})$. Thus $\mathcal{O}$ is a proper subset of the orthonormal set $y \cup \mathcal{O}$, which contradicts (*i*).

(*ii*) implies (*iii*). By Lemma I.7.13, $x - Pz \perp \overline{\text{sp}}\,(\mathcal{O})$. Since $\overline{\text{sp}}\,(\mathcal{O}) = \mathcal{H}$, $x - Px \perp x - Px$, whence $x = Px$.

(*iii*) implies (*iv*) trivially.

(*iv*) implies (*i*). Let $\mathcal{O}$ be a subset of orthonormal set $\mathcal{O}_1$. Assume there exists an $x \in \mathcal{O}_1$ but not in $\mathcal{O}$. Then $\langle x, u \rangle = 0$ for each $u \in \mathcal{O}$. Hence, by (*iv*), $1 = \|x\|^2 = \|Px\|^2 = \sum_{u \in \mathcal{O}} |\langle x, u \rangle|^2 = 0$. Therefore $\mathcal{O}_1 = \mathcal{O}$. Since $\mathcal{O}_1$ was an arbitrary orthonormal set containing $\mathcal{O}$, $\mathcal{O}$ is a maximal orthonormal set.

In the literature, a maximal orthonormal set in $\mathcal{H}$ is called *complete.*

In the $\mathcal{L}_2([0, 2\pi])$ space of real-valued functions, the orthonormal set $\{1/\sqrt{2\pi}, \cos nt/\sqrt{\pi}, \sin nt/\sqrt{\pi}, n = 1, 2, \ldots\}$ is a maximal orthonormal set. This statement is proved in McShane and Botts [1], pages 230–233.

I.7.23 Lemma. *In a separable Hilbert space, every orthonormal set is countable.*

Proof. Let $\mathcal{O}$ be an orthonormal set. For u and $v \in \mathcal{O}$

$$\|u - v\|^2 = \langle u - v, u - v\rangle = \|u\|^2 + \|v\|^2 = 2$$

Thus $\mathcal{O}$ is countable, since a set of isolated points in a separable metric space is countable.

I.7.24 Theorem. *A separable Hilbert space is equivalent to l_2.*

Proof. Given a separable Hilbert space $\mathcal{H}$, there exists, by Lemmas I.7.14 and I.7.23, a maximal orthonormal set $u_1, u_2, \ldots$ in $\mathcal{H}$. By Lemma I.7.22, each $x \in \mathcal{H}$ has the representation $x = \sum_{i=1}^{\infty} \langle x, u_i\rangle u_i$. Since

$$\|x\|^2 = \sum_{i=1}^{\infty} |\langle x, u_i\rangle|^2 \tag{1}$$

the sequence $\{\langle x, u_i\rangle\}$ is in l_2. Let T be the linear map from $\mathcal{H}$ into l_2 defined by

$$Tx = \{\langle x, u_i\rangle\}$$

By (1), T is an isometry. It remains to prove that $\mathcal{R}(T) = l_2$. Let $\{\alpha_i\} \in l_2$ be given. Defining $s_n = \sum_{i=1}^{n} \alpha_i u_i$, it follows that $\{s_n\}$ is a Cauchy sequence and therefore converges to some $v \in \mathcal{H}$. Hence

$$Tv = \left\{\left\langle \sum_{k=1}^{\infty} \alpha_k u_k, u_i\right\rangle\right\} = \{\alpha_i\}$$

chapter II

LINEAR OPERATORS AND THEIR CONJUGATES

There are essentially two methods of dealing with differential operators in the usual function space setting. The first is to define a new topology on the space so that the differential operators are continuous and then to develop and apply a general theory of continuous operators on a non-normable topological linear space. This is known as L. Schwartz's theory of distributions (cf. Schwartz [1], Hörmander [1], and Friedman [1]). The second method is to retain the Banach space structure while developing and applying a general theory of unbounded linear operators (cf. Browder [3] and Visik [1]). We use the second method for the following two reasons:

i. The linear differential operators usually encountered are closed or at least have closed linear extensions.
ii. Many of the important theorems which hold for continuous linear operators on a Banach space also hold for closed linear operators.

In this chapter, as well as in Chaps. IV and V, the second assertion is substantiated.

Throughout this chapter, X and Y are normed linear spaces over the same scalars, and T is a linear operator having domain a subspace of X

and range a subspace of Y. X and Y are assumed complete only when specifically stated.

For any set $M \subset X$, TM denotes the set $\{Tm \mid m \in M \cap \mathfrak{D}(T)\}$.

II.1 CLOSED LINEAR OPERATORS

II.1.1 Definition. *$X \times Y$ is defined as the normed linear space of all ordered pairs (x, y), $x \in X$, $y \in Y$, with the usual definitions of addition and scalar multiplication and with norm given by* $\|(x, y)\| = \max \{\|x\|, \|y\|\}$.

II.1.2 Definition. *The graph $G(T)$ of T is the set $\{(x, Tx) \mid x \in \mathfrak{D}(T)\}$. Since T is linear, $G(T)$ is a subspace of $X \times Y$.*

If the graph of T is closed in $X \times Y$, then T is said to be **closed** *in X. When there is no ambiguity concerning the space X, we say that T is closed.*

II.1.3 Remarks

i. T is closed if and only if $\{x_n\}$ in $\mathfrak{D}(T)$, $x_n \to x$, $Tx_n \to y$, imply $x \in \mathfrak{D}(T)$ and $Tx = y$.

ii. If T is 1-1 and closed, then T^{-1} is closed.

iii. The null manifold of a closed operator is closed.

iv. If $\mathfrak{D}(T)$ is closed and T is continuous, then T is closed.

v. The continuity of T does not necessarily imply that T is closed. Conversely, T closed does not necessarily imply that T is continuous. This statement can be verified by the following examples.

Let $\mathfrak{D}(T)$ be any proper dense subspace of $X = Y$ and let T be the identity map. T is obviously continuous but not closed.

II.1.4 Example. Let $X = Y = C([0, 1])$ and let $C'([0, 1)]$ be the subspace of X consisting of the functions with continuous first derivatives. Define the linear differential operator T mapping $C'([0, 1])$ into Y by $(Tx)(t) = x'(t)$, $t \in [0, 1]$. T is closed; for if $x_n \to x$ and $Tx_n \to y$, then $\{x_n\}$ converges uniformly to x and $\{x_n'\}$ converges uniformly to y on $[0, 1]$. It follows from taking antiderivatives of x_n' and y that x is in $C'([0, 1])$ and that $Tx = x' = y$ on $[0, 1]$. Thus T is closed. However, T is unbounded, since the sequence $\{x_n(t)\} = \{t^n\}$ has the properties $\|Tx_n\| = n$ and $\|x_n\| = 1$.

It is shown in Chaps. VI and VII that large classes of differential operators are closed.

The closed-graph theorem, which is another fundamental theorem in functional analysis, shows when a closed linear operator is continuous. While this theorem is valid in more general topological linear spaces (cf.

Robertson and Robertson [1]), we present the proof for Banach spaces as an application of basic lemma II.1.7. The lemma is also applied later to prove the important Theorem II.4.3.

II.1.5 Definition. *For Z a normed linear space and r a positive number, define*

$$S_Z(r) = \{z \mid z \in Z, \|z\| \leq r\} \qquad S_Z{}^0(r) = \{z \mid z \in Z, \|z\| < r\}$$

The following lemma is a simple consequence of the properties of a norm.

II.1.6 Lemma

i. *Given $x \in X$, the translation map f_x from X onto X defined by $f_x(z) = x + z$ is a homeomorphism.*
ii. *For any nonzero scalar a, the map g_a from X onto X, defined by $g_a(x) = ax$, is a homeomorphism.*
iii. *For any open set $V \subset X$ and any $x \in X$, $f_x(V) = x + V$ is open. Therefore $A + V = \bigcup_{x \in A} x + V$ is open, where A is an arbitrary set in X.*
iv. *Given a set $K \subset X$, $a\bar{K} = g_a(\bar{K}) = \overline{g_a(K)} = \overline{aK}$ for any scalar a.*

II.1.7 Basic Lemma. *Let X be complete and let T be closed. If $S_Y{}^0(r) \subset \overline{TS_X(1)}$, then $S_Y{}^0(r) \subset TS_X(1)$.*

Proof. To prove the lemma, it suffices to prove that $S_Y{}^0(r) \subset TS_X(1/1 - \varepsilon)$ whenever $0 < \varepsilon < 1$; for if this is the case, then given $y \in S_Y{}^0(r)$, there exists an ε, $0 < \varepsilon < 1$, such that $y/1 - \varepsilon$ is also in $S_Y{}^0(r)$. Hence there exists an $x \in S_X(1/1 - \varepsilon)$ such that $Tx = y/1 - \varepsilon$ or $T((1 - \varepsilon)x) = y$. Since $\|(1 - \varepsilon)x\| \leq 1$, y is in $TS_X(1)$. Let $0 < \varepsilon < 1$ and let $y \in S_Y{}^0(r)$ be given. By hypothesis and (*iv*) of Lemma II.1.6, it follows that for each nonnegative integer n, $S_Y{}^0(r\varepsilon^n) \subset \overline{TS_X(\varepsilon^n)}$. Taking $n = 0$, there exists an $x_0 \in S_X(1)$ such that $\|y - Tx_0\| < r\varepsilon$; that is, $y - Tx_0 \in S_Y{}^0(r\varepsilon)$. Hence, taking $n = 1$, there exists an $x_1 \in S_X(\varepsilon)$ such that $\|y - Tx_0 - Tx_1\| < r\varepsilon^2$; that is, $y - Tx_0 - Tx_1 \in S_Y{}^0(r\varepsilon^2)$. Proceeding in this manner, a sequence $\{x_n\}$ is obtained with the properties that

$$(1) \qquad x_n \in S_X(\varepsilon^n) \qquad \text{and} \qquad \|y - \sum_{i=0}^{n} Tx_i\| < r\varepsilon^{n+1}$$

Consequently, $\sum_{n=0}^{\infty} \|x_n\| \leq 1/1 - \varepsilon$ and therefore the sequence $\{z_n\}$ defined by $z_n = \sum_{i=0}^{n} x_i$ is a Cauchy sequence in Banach space X. Thus there exists an $x \in X$ such that $\|x\| \leq 1/1 - \varepsilon$ and $z_n \to x$. From (1) it is clear that $Tz_n \to y$. Since T is closed, x must be in $\mathfrak{D}(T) \cap S_X(1/1 - \varepsilon)$ and $Tx = y$, whence $y \in TS_X(1/1 - \varepsilon)$.

II.1.8 Open-mapping theorem. *Let X be complete and let Y be of the second category. If T is closed and $\mathfrak{R}(T) = Y$, then T is an open map; i.e., the map takes open sets onto open sets.*

Proof. To show that T is an open map, one need only prove that given $x \in X$ and $\varepsilon > 0$, the set $Tx + TS_X(\varepsilon)$ contains an open set $Tx + S_Y{}^0(r)$ for some $r > 0$. By the basic lemma II.1.7, it suffices to show the existence of some $r > 0$ such that $S_Y{}^0(r) \subset \overline{TS_X(\varepsilon)}$. Since $Y = \bigcup_{n=1}^{\infty} nTS_X(1)$ and Y is of the second category, there exists a positive integer p such that $p\,\overline{TS_X(1)} = \overline{pTS_X(1)}$ contains a nonempty open set. It follows from *(ii)* of Lemma II.1.6 that $\overline{TS_X(\varepsilon/2)}$ must also contain a nonempty open set V. Thus

$$0 \in V - V \subset \overline{TS_X\left(\frac{\varepsilon}{2}\right)} - \overline{TS_X\left(\frac{\varepsilon}{2}\right)} \subset \overline{TS_X(\varepsilon)}$$

Since $V - V$ is an open set about 0, there exists an $r > 0$ such that $S_Y{}^0(r) \subset V - V \subset \overline{TS_X(\varepsilon)}$. Thus the proof of the theorem is complete.

For an example of an incomplete normed linear space of the second category, the reader is referred to Bourbaki [1], Exercise 6, page 3.

II.1.9 Closed-graph theorem. *A closed linear operator mapping a Banach space into a Banach space is continuous.*

Proof. Suppose the domain of T is all of X with X and Y complete. The graph $G(T)$ may be considered as a Banach space, since it is a closed subspace of Banach space $X \times Y$. Define linear maps P_1 and P_2 from $G(T)$ into X and Y, respectively, by $P_1(x, Tx) = x$ and $P_2(x, Tx) = Tx$. Clearly, P_1 is 1-1 and satisfies the hypotheses of the open-mapping theorem. Consequently, P_1 has a continuous inverse. Since P_2 is continuous, $T = P_2P_1^{-1}$ is also continuous.

In the closed-graph theorem it is essential that both X and Y be complete, as may be seen in the following two examples.

The first-order differential operator in Example II.1.4 is closed but not continuous. The domain of the operator is $C'([0, 1])$, which is not complete, while the range of the operator is $Y = C([0, 1])$, which is complete.

II.1.10 Example. Let X be any infinite-dimensional Banach space and let H be a set of basis elements for X. H is usually referred to as a

Hamel basis. It may be assumed that the elements in H are of norm 1. Define X_1 as the vector space X with norm $\| \quad \|_1$ given by

$$\left\| \sum_{i=1}^{n} a_i h_i \right\|_1 = \sum_{i=1}^{n} |a_i| \qquad h_i \in H$$

Clearly, $\|x\|_1 \geq \|x\|$. X_1 is not complete. Let $y_1, y_2, \ldots$ be any infinite countable subset of H and let $s_n = \sum_{k=1}^{n} k^{-2} y_k$. It is easy to verify that $\{s_n\}$ is a Cauchy sequence in X_1 which does not converge in X_1. Let T be the identity map from X onto X_1. Then T is closed and has a continuous inverse. However, T is not continuous; otherwise T would be an isomorphism, which would imply that X_1 is complete.

As an application of the closed-graph theorem, the following fundamental theorem is obtained. It is also referred to in the literature as the Banach-Steinhaus theorem.

II.1.11 Uniform-boundedness principle. *Let X be a Banach space and let F be a set of continuous linear operators mapping X into Y such that for each $x \in X$, $\sup_{T \in F} \|Tx\| < \infty$. Then $\sup_{T \in F} \|T\| < \infty$; that is, the operators in F, when restricted to a bounded set in X, are uniformly bounded.*

Before proving the theorem, we indicate (at the expense of being somewhat verbose) how one might motivate the proof. It is required to find some constant M so that for all $x \in X$, $\sup_{T \in F} \|Tx\| \leq M\|x\|$. A technique which is often used is to fix x and thereby induce a vector-valued function $Ax: F \to Y$ defined by $(Ax)T = Tx$. In this context, we seek a constant M so that

$$\text{(1)} \qquad \sup_{T \in F} \|(Ax)T\| \leq M\|x\| \qquad x \in X$$

By hypothesis, $\sup_{T \in F} \|(Ax)T\| < \infty$, for each $x \in X$. Thus, Ax is a member of $B(F, Y)$, the normed linear space of bounded functions from F into Y with norm defined by $\|h\| = \sup_{T \in F} \|h(T)\|$. Hence, by (1), an M is required so that $\|Ax\| \leq M\|x\|$ for all $x \in X$. This, in turn, suggests defining the operator $A: X \to B(F, Y)$, with the hope that A is bounded. The proof of the theorem consists of showing that A is continuous by applying the closed-graph theorem. Since $B(F, Y)$ is not necessarily complete, $B(F, \hat{Y})$ is considered instead, where $\hat{Y}$ is the completion of Y.

Proof. Let A be the linear map from X into $B(F, \hat{Y})$ defined by $(Ax)T = Tx$, $T \in F$. It is easy to verify that A is closed, whence, by the closed-graph theorem, A is continuous. Therefore

$$\sup_{T \epsilon F} \|Tx\| = \|Ax\| \leq \|A\| \, \|x\| \qquad x \epsilon X$$

Consequently

$$\sup_{T \epsilon F} \|T\| \leq \|A\|$$

In the above theorem it is essential that X be complete. Let X be the subspace of l_2 consisting of all elements of the form $\sum_{i=1}^{k} a_i u_i$, where $u_1 = (1, 0, 0, \ldots)$, $u_2 = (0, 1, 0, 0, \ldots)$, etc. For each positive integer n, let $T_n \colon X \to l_2$ be the linear operator defined by setting

$$T_n u_i = \begin{cases} 0 & \text{if } i \neq n \\ n u_n & \text{if } i = n \end{cases}$$

Then $\|T_n\| = n$ while for each $x \epsilon X$, $T_n x \to 0$.

By taking Y to be the space of scalars in Theorem II.1.11, we obtain the result that if X is complete and F is a subset of X' such that $\sup_{x' \epsilon F} |x'x| < \infty$, $x \epsilon X$, then F is bounded in X'. The following theorem is what one might call the dual result.

II.1.12 Theorem. *Suppose K is a subset of X such that*

$$\sup_{k \epsilon K} |x'k| < \infty \qquad x' \epsilon X'$$

Then K is bounded.

Proof. Let J be the natural map from X into X''. By hypothesis,

$$\sup_{k \epsilon K} |(Jk)x'| = \sup_{k \epsilon K} |x'k| < \infty \qquad x' \epsilon X'$$

Since X' is complete and J is a linear isometry, the uniform-boundedness principle assures us that $\sup_{k \epsilon K} \|k\| = \sup_{k \epsilon K} \|Jk\| < \infty$.

II.1.13 Definition. *Let M be a subspace of X. An operator P is called a* **projection** *from X onto M if P is a bounded linear map from X onto M such that $P^2 = P$.*

As a further application of the closed-graph theorem we show that there is a 1-1 correspondence between the set of projections onto M and certain closed subspaces of X.

II.1.14 Theorem. *Let M be a closed subspace of Banach space X. There exists a projection from X onto M if and only if there exists a closed subspace N of X such that $X = M \oplus N$ ($M \cap N = (0)$ and $X = M + N$). In this case, there exists a $c > 0$ such that*

$$\|m + n\| \geq c\|m\| \qquad m \in M, n \in N$$

Proof. Suppose P is a projection from X onto M. Let $N = \mathfrak{N}(P)$. Since P is continuous, N is closed. Furthermore, $N \cap M = (0)$, since $v \in N \cap M$ implies $v = Pv = 0$. Since any $x \in X$ may be written $x = Px + (x - Px)$ and $x - Px$ is in N, $X = M + N$.

Conversely, suppose N satisfies the hypotheses of the theorem. Then each $x \in X$ has a unique representation of the form

$$x = m + n \qquad m \in M, n \in N$$

Define the operator P from X onto M by $P(m + n) = m$. Clearly, P is linear and $P^2 = P$. Moreover, P is closed. Indeed, suppose

$$x_k = m_k + n_k \rightarrow x \qquad m_k \in M, n_k \in N$$

$$Px_k = m_k \rightarrow y$$

Since M is closed, y is in M. Therefore $y = Py$ and $n_k \rightarrow x - y$. Since N is also closed, $x - y$ is in N. Hence $0 = P(x - y) = Px - y$. Thus P is closed. By the closed-graph theorem, P is continuous and

$$\|P\| \, \|m + n\| \geq \|P(m + n)\| = \|m\| \qquad m \in M, n \in N$$

The above proof shows that the 1-1 correspondence between projections P onto M and the closed subspaces N such that $X = M \oplus N$ is given by $P \rightarrow \mathfrak{N}(P)$.

II.1.15 Theorem. *There is always a projection from a Hilbert space onto any one of its closed subspaces.*

Proof. Corollary I.7.17 and Theorem II.1.14.

The above theorem is instrumental in obtaining a number of important results in Hilbert space. There are examples which show that the theorem does not hold in general if the Banach space is not required to be a Hilbert space. For a discussion of projections and corresponding references, the reader is referred to Dunford and Schwartz [1], pages 553 and 554, and to Pelczynski [1]. For finite-dimensional subspaces we have the following result.

II.1.16 Theorem. *There is always a projection from a normed linear space onto any one of its finite-dimensional subspaces.*

Proof. Suppose M is a finite-dimensional subspace of X with basis $x_1, x_2, \ldots, x_n$. Let $x'_1, x'_2, \ldots, x'_n$ be elements in X' such that $x'_i x_j = \delta_{ij}$, where δ_{ij} is the Kronecker delta. The x'_i may be constructed as follows. Let $M_i = \text{sp}\,\{x_1, x_2, \ldots, x_{i-1}, x_{i+1}, \ldots, x_n\}$. Since M_i is finite-dimensional and therefore closed, there exists a $v'_i \in X'$ such that $v'_i x_i \neq 0$ and $v'_i M_i = 0$. Choose $x'_i = v'_i / v'_i x_i$. It is easy to verify that the operator P defined by $Px = \sum_{i=1}^{n} x'_i(x) x_i$ is a projection from X onto M.

II.1.17 Remarks. If X is an n-dimensional normed linear space, then so is X'. To see this, let $x_1, \ldots, x_n$ and $x'_1, \ldots, x'_n$ be as in the proof of the above theorem, where now $X = M$. It is easy to verify that the set $\{x'_1, \ldots, x'_n\}$ is linearly independent and that each $x' \in X'$ has the representation $x' = \sum_{i=1}^{n} x'(x_i) x'_i$. Thus every finite-dimensional normed linear space is reflexive, since $J_X X$, X, X', and X'' have the same dimension and therefore $J_X X = X''$.

II.2 CONJUGATE OF A LINEAR OPERATOR

As we shall see in subsequent chapters, the concept of the conjugate of a linear operator is very useful in obtaining information about the range and inverse of T. In the remaining sections of this chapter, theorems are presented which show how certain properties of an operator and its conjugate are related.

When T is bounded on all of X, the conjugate of T is the map T' which takes each $y' \in Y'$ to $y'T \in X'$. For example, suppose $X = Y$ is an n-dimensional normed linear space over the complex numbers with basis $x_1, x_2, \ldots, x_n$. Let $M = (\alpha_{ij})$ be an $n \times n$ matrix whose elements are complex numbers. Then M, together with the basis $\{x_i\}$, determines T by setting $Tx_i = \sum_{j=1}^{n} \alpha_{ij} x_j$. Let $x'_1, x'_2, \ldots, x'_n$ be elements in X' such that $x'_i x_j = \delta_{ij}$, the Kronecker delta. Then T' is the operator determined by the adjoint matrix M^*, together with $\{x'_i\}$, defined by setting $T'x'_i = \sum_{j=1}^{n} \alpha_{ji} x'_j$.

Now if T is no longer continuous, then $y'T$ need not be in X'. To define in a similar manner the conjugate of such an operator, the natural procedure is to single out those $y' \in Y'$ for which $y'T$ is continuous. Letting this subspace of Y' be the domain $\mathfrak{D}(T')$ of T', define $T' : \mathfrak{D}(T') \to X'$ by $T'y' = y'T$. The requirement that $\mathfrak{D}(T) = X$ is also too restrictive, especially in cases when T is a differential operator whose domain, considered as a subspace of $\mathcal{L}_2(\Omega)$, consists of certain smooth functions. In most applications, $\mathfrak{D}(T) \neq X$ but $\overline{\mathfrak{D}(T)} = X$, where X is complete.

Before defining T' formally, we prove the following theorem.

II.2.1 Theorem. *Let M be a subspace dense in X and let Y be complete. If A is a bounded linear map from M into Y, then there exists a unique continuous linear extension $\bar{A}$ of A to all of X and $\|A\| = \|\bar{A}\|$. The conjugate space M' is equivalent to X'.*

Proof. Given $x \in X$, there exists a sequence $\{x_k\}$ in M which converges to x. Since $\|Ax_k - Ax_j\| \leq \|A\| \, \|x_k - x_j\|$, $\{Ax_k\}$ is a Cauchy sequence which therefore converges to some y in Banach space Y. Let $\bar{A}x = y$. In order to show that $\bar{A}$ is unambiguously defined, suppose that $\{z_k\}$ is a sequence in M which also converges to x. Then by what has just been shown, the sequence $Ax_1, Az_1, Ax_2, Az_2, \ldots$ converges to some $w \in Y$. Thus subsequences $\{Ax_k\}$ and $\{Az_k\}$ both converge to w. Since $Ax_k \to y$, $y = w = \lim_{k \to \infty} Az_k$. Clearly, $\bar{A}$ is a linear extension of A to all of X. Now

$$\|\bar{A}x\| = \lim_{n \to \infty} \|Ax_n\| \leq \|A\| \lim_{n \to \infty} \|x_n\| = \|A\| \, \|x\|$$

Hence $\|\bar{A}\| \leq \|A\|$. Since $\bar{A}$ is an extension of A, $\|\bar{A}\| \geq \|A\|$. Thus $\|A\| = \|\bar{A}\|$. If Y is taken to be the scalars, then the map from M' onto X' which takes A in M' to $\bar{A} \in X'$ is a linear isometry.

II.2.2 Definition. *Let the domain of T be dense in X. The* ***conjugate*** *T' of T is defined as follows.*

$\mathfrak{D}(T') = \{y' \mid y' \in Y', \ y'T$ continuous on $\mathfrak{D}(T)\}$. For $y' \in \mathfrak{D}(T')$, let T' be the operator which takes $y' \in \mathfrak{D}(T')$ to $\overline{y'T}$, where $\overline{y'T}$ is the unique continuous linear extension of $y'T$ to all of X.

$\mathfrak{D}(T')$ is a subspace of Y', and T' is linear.

$T'y'$ is taken to be $\overline{y'T}$ rather than $y'T$ in order that $\mathfrak{R}(T')$ be contained in X'.

The requirement that the domain of the operator be dense in X is not as restrictive as might appear at first glance. Suppose $\mathfrak{D}(T)$ is not dense in X. Then one can define X_1 as the closure of the domain of T and obtain the conjugate T' as a mapping from a subspace of Y' into X_1'. By inspecting the theorems which require that $\overline{\mathfrak{D}(T)} = X$, one can usually remove this restriction by considering X_1 in place of X. This is done, for example, in the proofs of Theorems V.1.6 and V.1.8.

The following simple examples are presented in detail in order to give some "feeling" for the definition of a conjugate operator. The conjugates of certain differential operators are determined in Chap. VI.

II.2.3 Example. Let $X = Y = l_p$, $1 \leq p < \infty$, and let

$$u_1 = (1, 0, 0, \ldots), \; u_2 = (0, 1, 0, \ldots), \text{ etc.}$$

be the unit vectors in l_p. Define T by

$$\mathfrak{D}(T) = \text{sp }\{u_k\}$$

$$T(x_1, x_2, \ldots, x_n, 0, 0, \ldots) = \left(\sum_{j=1}^{n} jx_j, x_2, x_3, \ldots, x_n, 0, 0, \ldots\right)$$

Suppose $y' = (a_1, a_2, \ldots) \in \mathfrak{D}(T')$. Then for $k \geq 1$,

$$|y'Tu_k| = |a_1k + a_k| \geq |a_1|k - |a_k| \geq |a_1|k - \|y'\|$$

Since $\|u_k\| = 1$ and $y'T$ is bounded on $\mathfrak{D}(T)$, $a_1 = 0$. Also, any element $(0, b_1, b_2, \ldots) \in l_{p'} = l_p'$ is in $\mathfrak{D}(T')$. Hence the domain of T' consists of all the elements in $l_{p'}$ which have zero as their first term. Suppose $T'y' = (c_1, c_2, \ldots)$, where $y' = (0, a_2, a_3, \ldots) \in \mathfrak{D}(T')$. Then

$$c_k = T'y'u_k = y'Tu_k = a_k, \qquad k \geq 2$$

and $c_1 = 0$. Thus $T'y' = y'$.

II.2.4 Example. Let $X = Y = \mathfrak{L}_2([0, 1])$ and let $\mathfrak{D}(T)$ be the subspace of X consisting of those f such that

(1) f is absolutely continuous on $[0, 1]$ with $f' = \dfrac{df}{dt} \in \mathfrak{L}_2([0, 1])$

(2) $$f(0) = f(1) = 0$$

Define T by $Tf = f'$. Strictly speaking, $\mathfrak{D}(T)$ is the subspace of cosets $[f] \in X$, where f satisfies (1) and (2) and Tf is the coset $[f']$. Recall that an absolutely continuous function is differentiable almost everywhere. Since $C_0^\infty(I)$ is contained in $\mathfrak{D}(T)$, $\mathfrak{D}(T)$ is dense in X by Theorem 0.9. Suppose $y \in \mathfrak{D}(T') \subset Y' = \mathfrak{L}_2([0, 1])$ and $T'y = x$. Then for $f \in \mathfrak{D}(T)$,

(3) $$\int_0^1 y(t)f'(t)\,dt = yTf = (T'y)f = \int_0^1 x(t)f(t)\,dt$$

Since f satisfies (1) and (2) and x is in $\mathfrak{L}_2([0, 1]) \subset \mathfrak{L}_1([0, 1])$, we may integrate by parts and get

(4) $$\int_0^1 x(t)f(t)\,dt = -\int_0^1 (H(t) + C)f'(t)\,dt$$

where C is an arbitrary constant and $H(t) = \int_0^t x(s)\,ds$. Thus, by (3) and (4),

$$0 = \int_0^1 (y(t) + H(t) + C)f'(t)\,dt \qquad f \in \mathfrak{D}(T) \tag{5}$$

Let $f_0(t) = \int_0^t \overline{(y(s) + H(s) + C_0)}\,ds$, where C_0 is chosen so that $f_0(1) = 0$. Then f_0 is in $\mathfrak{D}(T)$, and from (5) it follows that

$$0 = \int_0^1 |y(t) + H(t) + C_0|^2\,dt$$

Hence, as elements in X, $y = -H - C_0$ (considered as functions, $y = -H - C_0$ a.e.). It follows from the definition of H that y satisfies (1) and $y' = -x = -T'y$. On the other hand, if y satisfies (1), then considering y as an element of Y', Hölder's inequality and integration by parts give

$$|yTf| = \left|\int_0^1 y(t)f'(t)\,dt\right| = \left|\int_0^1 y'(t)f(t)\,dt\right| \leq \|y'\|\,\|f\| \qquad f \in \mathfrak{D}(T)$$

Thus y is in $\mathfrak{D}(T')$, since $\|yT\| \leq \|y'\|$. We have shown that $\mathfrak{D}(T')$ is the subspace of $\mathfrak{L}_2([0, 1])$ consisting of those elements y which satisfy (1) and for which $T'y = -y'$. Hence T' is a proper extension of $-T$.

II.2.5 Example. Let $X = \mathfrak{L}_p(I)$, $Y = \mathfrak{L}_q(I)$, $1 \leq p, q < \infty$, where I is a compact interval. Let k be a bounded measurable function on $I \times I$. Define the linear map K from X into Y by

$$(Kf)(t) = \int_I k(s, t)f(s)\,ds$$

It follows from Hölder's inequality that K is in $[X, Y]$. Suppose $y \in Y' = \mathfrak{L}_{q'}(I)$ and $K'y = x \in X' = \mathfrak{L}_{p'}(I)$. Then for $f \in X$,

$$\int_I x(s)f(s)\,ds = (K'y)f = yKf = \int_I y(t)\left[\int_I k(s, t)f(s)\,ds\right]dt$$

Hence, by Fubini's theorem,

$$\int_I x(s)f(s)\,ds = \int_I f(s)\left[\int_I k(s, t)y(t)\,dt\right]ds \qquad f \in X$$

Thus, we may infer from Theorem I.6.2 that

$$(K'y)(s) = x(s) = \int_I k(s, t)y(t)\,dt \qquad y \in Y'$$

Unless otherwise specified, we shall assume throughout the remainder of this chapter that the domain of T is dense in X.

II.2.6 Theorem. *T' is a closed linear operator in Y'.*

Proof. Suppose $y'_n \to y'$ and $T'y'_n \to x'$. Then for each $x \in \mathfrak{D}(T)$, $y'_n Tx \to y'Tx$ and $y'_n Tx = T'y'_n x \to x'x$. Thus $y'T = x'$ on $\mathfrak{D}(T)$. Hence, by the definition of T', $y' \in \mathfrak{D}(T')$ and $T'y' = x'$. Therefore T' is closed.

The above theorem is used in Chap. VI to prove that certain differential operators are closed. For instance, the operator T_1 defined by

$$\begin{aligned} \mathfrak{D}(T_1) &= \{f \mid f \text{ absolutely continuous on } [0, 1],\ f' \in \mathfrak{L}_2([0, 1])\} \\ T_1 f &= f' \end{aligned}$$

is closed in $\mathfrak{L}_2([0, 1])$, since it was shown that $-T_1$ is the conjugate of the operator T defined in Example II.2.4.

Next, some topological properties as well as the "size" of $\mathfrak{D}(T')$ are investigated.

Even though $\mathfrak{D}(T)$ may be all of X, it is still possible for $\mathfrak{D}(T')$ to consist solely of the zero vector. The following example which demonstrates this is due to Berberian.

II.2.7 Example. Take $X = Y = l_2$ and take $\mathfrak{D}(T)$ to be the span of the unit vectors $u_k = (0, 0, \ldots, \underset{k}{1}, 0, 0, \ldots) \in l_2$. Let

$$\{u_{kj} \mid k, j = 1, 2, \ldots\}$$

be any double indexing of $\{u_k\}$. For each k, define

$$Tu_k = u_k, \qquad j = 1, 2, \ldots$$

and extend T linearly to X. Suppose $y' = (a_1, a_2, \ldots) \in \mathfrak{D}(T')$. Then for each k, $T'y'u_{kj} = y'u_k = a_k$, $j = 1, 2, \ldots$. Now

$$\sum_{j=1}^{\infty} |T'y'u_{kj}|^2 \leq \|T'y'\|^2$$

Hence

$$0 = \lim_{j \to \infty} T'y'u_{kj} = a_k \qquad 1 \leq k$$

Therefore $y' = 0$.

For an example of a differential operator T with $\mathfrak{D}(T') = (0)$, the reader is referred to Stone [1], Theorem 10.10, page 447.

II.2.8 Theorem. *$\mathfrak{D}(T') = Y'$ if and only if T is continuous. If that is the case, then T' is also continuous and $\|T'\| = \|T\|$.*

Proof. Clearly, if T is continuous, then $y'T$ is continuous for each $y' \in Y'$. Thus $\mathfrak{D}(T') = Y'$. Suppose $\mathfrak{D}(T') = Y'$. Let S be the 1-sphere of $\mathfrak{D}(T)$. For each $y' \in Y'$, $\sup_{x \in S} |y'Tx| \leq \|T'y'\|$. Hence, by Theorem II.1.12, $\|T\| = \sup_{x \in S} \|Tx\| < \infty$. Now, for each $x \in S$, $|T'y'x| \leq \|y'\| \, \|T\|$. Thus $\|T'y'\| \leq \|T\| \, \|y'\|$, and therefore $\|T'\| \leq \|T\|$. By Corollary I.5.7,

$$\|Tx\| = \sup_{\|y'\|=1} |y'Tx| = \sup_{\|y'\|=1} |T'y'x| \leq \|T'\| \, \|x\| \qquad x \in \mathfrak{D}(T)$$

Hence $\|T\| \leq \|T'\|$ and the theorem follows.

The linear operators which one usually encounters have the property that they are restrictions of closed operators. Theorem II.2.11 characterizes such operators.

II.2.9 Definition. *A set F of linear functionals on a vector space V is said to be **total** if given any $v \neq 0$ in V, there exists an $f \in F$ such that $f(v) \neq 0$.*

In Examples II.2.3 and II.2.7, the domain of T' is not total. Corollary I.5.6 shows that a conjugate space is total.

II.2.10 Definition. *T is called **closable** if there exists a linear extension of T which is closed in X. The domain of T is not required to be dense in X.*

II.2.11 Theorem. *Statements (i), (ii) and (iii) are equivalent. ($\mathfrak{D}(T)$ is not required to be dense in X.)*

i. T is closable.
ii. T has a minimal closed linear extension; i.e., there exists a closed linear extension $\bar{T}$ of T such that any closed linear extension of T is a closed linear extension of $\bar{T}$.
iii. For any $y \neq 0$ in Y, $(0, y)$ is not in the closure of the graph of T.

If $\overline{\mathfrak{D}(T)} = X$, then the statement that $\mathfrak{D}(T')$ is total is equivalent to the above three statements. In this case, $T' = (\bar{T})'$, where $\bar{T}$ is the minimal closed extension of T.

Proof. *(i) implies (iii).* Let $\tilde{T}$ be a closed linear extension of T. If $y \in Y$ and $y \neq 0$, then $(0, y) \notin G(\tilde{T}) \supset G(T)$. Hence $(0, y) \notin \overline{G(T)}$, since $G(\tilde{T})$ is closed in $X \times Y$.

(iii) implies (ii). Suppose $(0, y) \notin \overline{G(T)}$ for any $y \neq 0$ in Y. Define $\bar{T}$ as the operator whose graph is $\overline{G(T)}$; that is,

$$\mathfrak{D}(\bar{T}) = \{x \mid (x, z) \in \overline{G(T)} \text{ for some } z \in Y\} \qquad \bar{T}x = z$$

Then $\bar{T}$ is unambiguously defined and is a closed linear extension of T. Furthermore, $\bar{T}$ is the minimal closed linear extension of T; for if T_1 is a closed linear extension of T, then $G(T_1) \supset \overline{G(T)} = G(\bar{T})$.

(ii) implies (i) trivially.

Suppose that $\overline{\mathfrak{D}(T)} = X$. We proceed to show that $\mathfrak{D}(T')$ is total if and only if statement *(iii)* is valid. Let $\mathfrak{D}(T')$ be total and let $(0, y)$ be in $\overline{G(T)}$. Then there exists a sequence $\{x_n\}$ in $\mathfrak{D}(T)$ such that $x_n \to 0$ and $Tx_n \to y$. Thus, for each $y' \in \mathfrak{D}(T')$, $y'Tx_n \to 0$ and $y'Tx_n \to y'y$. Since $\mathfrak{D}(T')$ is total, it follows that $y = 0$.

Assume (iii). Let $y \neq 0$ be an element in Y. Then $(0, y) \notin \overline{G(T)}$ and therefore there exists a $z' \in (X \times Y)'$ such that $z'(0, y) \neq 0$ and $z'\overline{G(T)} = 0$. Defining $x' \in X'$ and $y' \in Y'$ by $x'x = z'(x, 0)$ and $y'y = z'(0, y)$, we obtain

$$0 = z'(x, Tx) = x'x + y'Tx \qquad x \in \mathfrak{D}(T)$$

$$0 \neq z'(0, y) = y'y$$

From these two equations, we have $y' \in \mathfrak{D}(T')$ and $y'y \neq 0$. Thus $\mathfrak{D}(T')$ is total.

Suppose $y' \in \mathfrak{D}((\bar{T})')$. Then $y'\bar{T}$ is continuous on $\mathfrak{D}(\bar{T})$ and, in particular, continuous on $\mathfrak{D}(T)$. Thus $y' \in \mathfrak{D}(T')$. Suppose $y' \in \mathfrak{D}(T')$ and $x \in \mathfrak{D}(\bar{T})$. Since $(x, \bar{T}x) \in G(\bar{T}) = \overline{G(T)}$, there exists a sequence $\{x_n\}$ in $\mathfrak{D}(T)$ such that $x_n \to x$ and $Tx_n \to \bar{T}x$. Hence

$$|y'\bar{T}x| = \lim_{n\to\infty} |y'Tx_n| \leq \|T'y'\| \lim_{n\to\infty} \|x_n\| = \|T'y'\| \, \|x\|$$

Therefore $y'\bar{T}$ is bounded, which means $y' \in \mathfrak{D}((\bar{T})')$. Hence

$$\mathfrak{D}(T') = \mathfrak{D}((\bar{T})')$$

Since $T'y' = (\bar{T})'y'$ on the dense subspace $\mathfrak{D}(T)$, it follows that $T' = (\bar{T})'$.

The theorem just proved shows that the operator in Example II.2.3 is not closable.

***II.2.12* Corollary.** *A linear operator which maps a Banach space into a Banach space is continuous if and only if the domain of its conjugate operator is total.*

Proof. Let X and Y be complete and let $\mathfrak{D}(T) = X$. Suppose $\mathfrak{D}(T')$ is total. Then by Theorem II.2.11, T is closable. Since $\mathfrak{D}(T) = X$, T must be its own closed extension. Hence T is continuous by the closed-graph theorem. If T is continuous, then $\mathfrak{D}(T') = Y'$, which is total.

II.2.13 Lemma. *If X is reflexive and F is a subspace of X', then F is total if and only if F is dense in X'.*

Proof. If F is total but not dense in X', there exists an $x'' \neq 0$ in X'' such that $x''F = 0$. By the reflexivity of X, there exists an $x \neq 0$ such that $0 = x''x' = x'x$ for all $x' \in F$. But this is impossible since F is total. If F is dense in X', then F is total, since any set dense in a total set is also total.

II.2.14 Theorem. *Suppose Y is reflexive. Then T is closable if and only if $\mathfrak{D}(T')$ is dense in Y'. In that case, the minimal closed extension of T is $J_Y{}^{-1}T''J_X$, where J_X and J_Y are the natural maps from X into X'' and Y onto Y'', respectively.*

Proof. The first assertion of the theorem is a consequence of Lemma II.2.13 and Theorem II.2.11. Thus the conjugate T'' of T' is defined, and by Theorem II.2.6, T'' is closed. Since J_Y and J_X are isomorphisms, it follows that $E = J_Y{}^{-1}T''J_X$ is also closed. Given $x \in \mathfrak{D}(T''J_X)$ and $y' \in \mathfrak{D}(T')$,

$$y'(J_Y{}^{-1}T''J_Xx) = (T''J_Xx)y' = (J_Xx)T'y' = T'y'x \tag{1}$$

Now $\mathfrak{D}(T) \subset \mathfrak{D}(T''J_X)$; for if $x \in \mathfrak{D}(T)$, then

$$|(J_Xx)T'y'| = |T'y'x| \leq \|y'\|\,\|Tx\| \qquad y' \in \mathfrak{D}(T')$$

Thus $\|(J_Xx)T'\| \leq \|Tx\|$, and therefore, by the definition of $\mathfrak{D}(T')$, $J_Xx \in \mathfrak{D}(T'')$. Consequently, we obtain from (1)

$$y'(Ex) = T'y'x = y'Tx \qquad x \in \mathfrak{D}(T),\ y' \in \mathfrak{D}(T')$$

Since $\mathfrak{D}(T')$ is total, $Ex = Tx$ for all $x \in \mathfrak{D}(T)$. Thus E is a closed linear extension of T. To prove that E is the minimal closed linear extension of T, it suffices to prove that $G(E) \subset \overline{G(T)} = G(\bar{T})$. Suppose $(u, v) \in G(E)$ but $(u, v) \notin \overline{G(T)}$. Then there exists a $z' \in (X \times Y)'$ such that

$$z'(u, v) \neq 0$$

$$z'(x, Tx) = 0 \qquad x \in \mathfrak{D}(T)$$

Let $x' \in X'$ and $y' \in Y'$ be defined by

$$x'x = z'(x, 0) \qquad y'y = z'(0, y)$$

Then

$$x'u + y'v = z'(u, v) \neq 0 \tag{2}$$

$$x'x + y'Tx = z'(x, Tx) = 0 \qquad x \in \mathfrak{D}(T) \tag{3}$$

Equation (3) implies $y' \in \mathfrak{D}(T')$ and $T'y' = -x'$. Since

$$v = Eu = J_Y^{-1}T''J_Xu$$

a substitution of u for x in (1) gives

$$y'v = T'y'u = -x'u$$

which contradicts (2). Hence $G(E) \subset \overline{G(T)}$.

II.2.15 Theorem. *T' is continuous if and only if $\mathfrak{D}(T')$ is closed in Y'.*

Proof. Suppose T' is continuous and $y_n' \to y'$, $y_n' \in \mathfrak{D}(T')$. Since T' and $\{y_n'\}$ are bounded, there exists some constant M such that $\|T'y_n'\| \leq M$. Therefore

$$|y'Tx| = \lim_{n\to\infty} |y_n'Tx| = \lim_{n\to\infty} |T'y_n'x| \leq M\|x\|$$

As a result, $y' \in \mathfrak{D}(T')$ whence $\mathfrak{D}(T')$ is closed.

Conversely, if $\mathfrak{D}(T')$ is closed, then T' is a closed linear map from Banach space $\mathfrak{D}(T')$ into Banach space X'. By the closed-graph theorem, T' is continuous.

Under the assumptions that X and Y are complete and that T is closed, it is shown in Corollary II.4.8 that the statements "T continuous," "T' continuous," and "$\mathfrak{D}(T')$ closed" are all equivalent.

II.2.16 Theorem. *Every operator in $[Y', X']$ is a conjugate of an operator in $[X, Y]$ if and only if Y is reflexive.*

Proof. Suppose Y is reflexive and $A \in [Y', X']$. Define $T \in [X, Y]$ by $T = J_Y^{-1}A'J_X$. Then $T' = A$, since for all $y' \in Y'$ and $x \in X$,

$$T'y'x = y'(J_Y^{-1}A'J_Xx) = (A'J_Xx)y' = (J_Xx)Ay' = Ay'x$$

Suppose every operator in $[Y', X']$ is a conjugate operator. Let $x_0' \in X'$ and $x_0 \in X$ be chosen so that $x_0'x_0 = 1$. Given $y'' \in Y''$, define $A \in [Y', X']$ by

$$Ay' = y''(y')x_0'$$

By hypothesis, there exists a $T \in [X, Y]$ such that $A = T'$. Hence,

$$y'Tx_0 = T'y'x_0 = Ay'x_0 = y''(y')x_0'x_0 = y''y' \qquad y' \in Y'$$

Therefore Y is reflexive.

II.3 STATES OF LINEAR OPERATORS

We shall first explain the notion of a state diagram. The diagram is a bookkeeping device for keeping track of some theorems concerning the range and inverse of T as well as T'.

Motivated by the known theorems relating T and T', we classify various possibilities of $\mathfrak{R}(T)$ and T^{-1} by

I: $\mathfrak{R}(T) = Y$
II: $\mathfrak{R}(T) \neq Y$ but $\overline{\mathfrak{R}(T)} = Y$
III: $\overline{\mathfrak{R}(T)} \neq Y$

1: T^{-1} exists and is continuous.
2: T^{-1} exists but is not continuous.
3: T has no inverse.

If $\mathfrak{R}(T) = Y$, we say that T is in state I or that T is *surjective*, written $T \in$ I. Similarly, we say that T is in state 3, written $T \in 3$, if T has no inverse. If $T \in$ II and $T \in 1$, we write $T \in \text{II}_1$. The same notation is used for T'. If $T \in \text{III}_1$ and $T' \in \text{I}_3$, we write $(T, T') \in (\text{III}_1, \text{I}_3)$. Thus, (T, T') has 81 classifications, which are described in the "checkerboard" II.3.14, which is referred to as a *state diagram*.

In this section, theorems are proved which show the impossibility of certain states for (T, T'). These eliminated states are shown by shading the corresponding squares. Some of the squares can be eliminated if additional assumptions are put on X and Y, for example, X reflexive or Y complete. In these instances the letters X-R or Y are put in the corresponding squares. The state diagrams II.3.14 and II.4.11 were obtained by Goldberg [1].

II.3.1 Theorem. *If T' has a bounded inverse, then $\mathfrak{R}(T')$ is closed.*

Proof. Suppose $T'y_n' \to x' \in X'$. Since T' has a bounded inverse, there exists an $m > 0$ such that

$$\|T'y_n' - T'y_k'\| \geq m\|y_n' - y_k'\|$$

Thus $\{y_n'\}$ is a Cauchy sequence which converges to some y' in Banach space Y'. Since T' is closed, y' is in $\mathfrak{D}(T')$ and $T'y' = x'$. Hence $\mathfrak{R}(T')$ is closed.

The theorem shows that state II_1 is impossible for T'. For brevity, we write $T' \notin II_1$.

II.3.2 Definition. *If C is a subset of X', then the orthogonal complement of C in X is the set ${}^\perp C = \{x \mid x \in X, x'x = 0$ for all $x' \in C\}$.*

II.3.3 Remarks. $K^\perp$ and ${}^\perp C$ are closed subspaces of X' and X, respectively. Also $K^\perp = \bar{K}^\perp$ and ${}^\perp C = {}^\perp \bar{C}$.

The proofs of Theorems II.3.4 to II.3.8 are left to the reader.

II.3.4 Theorem. *If M is a subspace of X, then ${}^\perp(M^\perp) = \bar{M}$.*

II.3.5 Theorem. *If N is a subspace of X', then $({}^\perp N)^\perp \supset \bar{N}$. If X is reflexive, then $({}^\perp N)^\perp = \bar{N}$.*

II.3.6 Remarks. Dieudonné [1] has shown that if N is a reflexive subspace of X', then $({}^\perp N)^\perp = N$. Since we later use the fact that $({}^\perp N)^\perp = N$, for N finite-dimensional, we shall prove this special case of Dieudonné's theorem.

Since $N \subset ({}^\perp N)^\perp$, it follows from Theorem I.6.4 that

$$\dim \frac{X}{{}^\perp N} = \dim \left(\frac{X}{{}^\perp N}\right)' = \dim ({}^\perp N)^\perp \geq \dim N \tag{1}$$

If $x_1', x_2', \ldots, x_n'$ is a basis for N, then the map A from $X/{}^\perp N$ into U^n defined by $A[x] = (x_1'x, x_2'x, \ldots, x_n'x)$ is 1-1 and linear. Thus

$$\dim \frac{X}{{}^\perp N} \leq n = \dim N \tag{2}$$

It follows from (1) and (2) that $N = ({}^\perp N)^\perp$.

II.3.7 Theorem

i. $\overline{\mathfrak{R}(T)}^\perp = \mathfrak{R}(T)^\perp = \mathfrak{N}(T')$

ii. $\overline{\mathfrak{R}(T)} = {}^\perp\mathfrak{N}(T')$

In particular, T has a dense range if and only if T' is 1-1.

By interchanging the roles of T and T' in the above theorem, we do not quite obtain the dual type theorems.

II.3.8 Theorem

i. ${}^\perp\mathfrak{R}(T') \supset \mathfrak{N}(T)$

ii. If $\mathfrak{D}(T')$ is total, then $\mathfrak{N}(T) = {}^\perp\mathfrak{R}(T') \cap \mathfrak{D}(T)$

iii. $\overline{\mathfrak{R}(T')} \subset \mathfrak{N}(T)^\perp$

In particular, if $\mathfrak{R}(T')$ is total, then T is 1-1.

II.3.9 Theorem. *If T and T' each has an inverse, then $(T^{-1})' = (T')^{-1}$.*

Proof. By Theorem II.3.7, $\mathfrak{D}(T^{-1}) = \mathfrak{R}(T)$ is dense in Y. Hence $(T^{-1})'$ is defined. Suppose $x' \in \mathfrak{D}((T')^{-1}) = \mathfrak{R}(T')$. Then there exists a $y' \in \mathfrak{D}(T')$ such that $T'y' = x'$. To show $x' \in \mathfrak{D}((T^{-1})')$, it suffices to prove that $x'T^{-1}$ is continuous on $\mathfrak{R}(T)$. This is certainly the case since

$$x'T^{-1}(Tx) = T'y'x = y'Tx \qquad x \in \mathfrak{D}(T)$$

Thus $(T^{-1})'x' = y'$ on $\mathfrak{R}(T)$, whence $y' = (T^{-1})'x' = (T^{-1})'T'y'$ since $\mathfrak{R}(T)$ is dense in Y. This shows that $(T^{-1})' = (T')^{-1}$ on $\mathfrak{D}((T')^{-1})$. It remains to prove $\mathfrak{D}((T^{-1})') \subset \mathfrak{D}((T')^{-1})$.

Suppose $z' \in \mathfrak{D}((T^{-1})')$. To show $z' \in \mathfrak{D}((T')^{-1}) = \mathfrak{R}(T')$, an element $v' \in \mathfrak{D}(T')$ will be exhibited so that $T'v' = z'$ or, equivalently, $v'T = z'$ on $\mathfrak{D}(T)$. It is clear that we should define v' as the continuous linear extension of $z'T^{-1}$ to all of Y, thereby obtaining $T'v' = z'$. Thus

$$\mathfrak{D}((T^{-1})') \subset \mathfrak{D}((T')^{-1})$$

II.3.10 Lemma. *If T does not have a bounded inverse, there exists a sequence $\{x_n\}$ in $\mathfrak{D}(T)$ such that $\|x_n\| \to \infty$ and $Tx_n \to 0$.*

Proof. There exists a sequence $\{z_n\}$ in $\mathfrak{D}(T)$ that $\|z_n\| = 1$ and $Tz_n \to 0$. Define

$$x_n = \begin{cases} \dfrac{z_n}{\|Tz_n\|^{\frac{1}{2}}} & \text{if } Tz_n \neq 0 \\ nz_n & \text{if } Tz_n = 0 \end{cases}$$

II.3.11 Theorem. *$\mathfrak{R}(T') = X'$ if and only if T has a bounded inverse.*

Proof. Suppose $\mathfrak{R}(T') = X'$. If T does not have a bounded inverse, there exists a sequence $\{x_n\}$ in $\mathfrak{D}(T)$ such that $\|x_n\| \to \infty$ while $Tx_n \to 0$. Thus, for each $y' \in \mathfrak{D}(T')$, $T'y'x_n \to 0$ whence $x'x_n \to 0$ for each $x' \in X'$. But then, by Theorem II.1.12, $\{x_n\}$ is bounded, which is a contradiction.

Conversely, suppose T has a bounded inverse. For $x' \in X'$, $x'T^{-1}$ is continuous on $\mathfrak{R}(T)$. Let y' be any continuous linear extension of $x'T^{-1}$ to all of Y. Then $y'T = x'$ on $\mathfrak{D}(T)$, whence $y' \in \mathfrak{D}(T')$ and $T'y' = x'$. Since $x' \in X'$ was arbitrary, $\mathfrak{R}(T') = X'$.

II.3.12 Theorem. *$\overline{\mathfrak{R}(T)} = Y$ and T has a bounded inverse if and only if $\mathfrak{R}(T') = X'$ and T' has a bounded inverse.*

Proof. By Theorems II.3.7 and II.3.11, $\overline{\mathfrak{R}(T)} = Y$ and T has a bounded inverse if and only if T' has an inverse defined on all of X'. Since T' is closed, so is $(T')^{-1}$. Hence, by the closed-graph theorem, $\mathfrak{D}((T')^{-1}) = X'$ if and only if $(T')^{-1}$ is continuous on X'.

The next theorem shows that additional states for (T, T') fail to exist when Y is complete.

II.3.13 Theorem. *Let Y be complete. If $\mathcal{R}(T) = Y$, then T' has a bounded inverse.*

Proof. The argument is analogous to the one in Theorem II.3.11, with use being made of Theorem II.1.11.

As a summary of the preceding theorems, the following state diagram is obtained. For example, we see from an inspection of the diagram that T has a bounded inverse if and only if the range of T' is all of X'.

In regard to the squares which remain open, we shall present in Sec. II.5 examples which exhibit the existence of the corresponding states when both X and Y are reflexive.

II.3.14 State diagram for linear operators

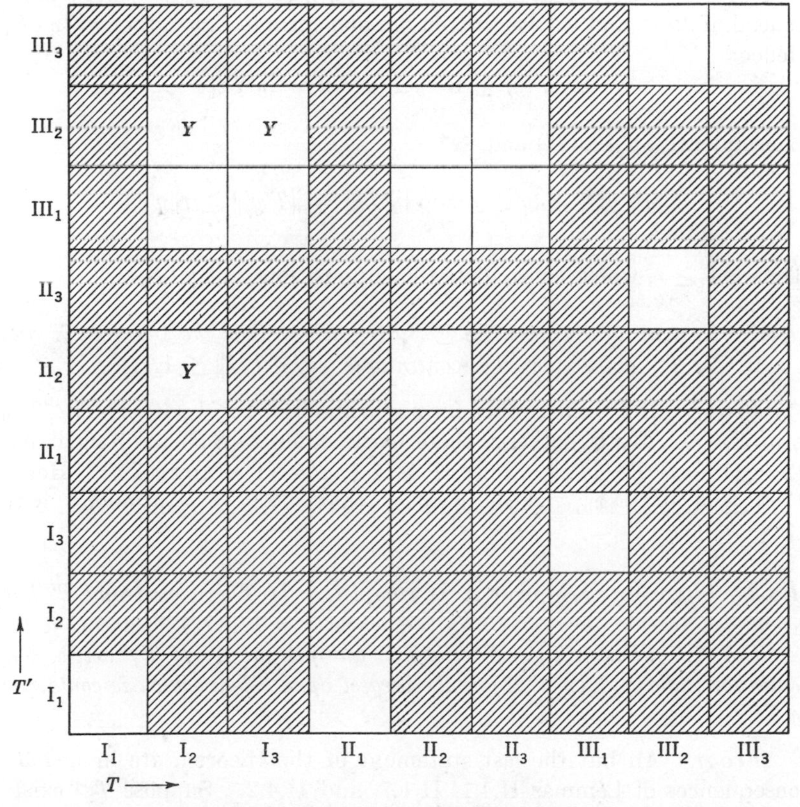

Y: Cannot occur if Y is complete.

II.4 STATES OF CLOSED LINEAR OPERATORS

This section is devoted to the construction of a state diagram for closed operators. Actually, the only "deep" theorem is II.4.3, which plays a very important role in later sections. The theorem was first proved for bounded, instead of closed, operators by Banach [1], Theorem 1, page 146. For the generalization to closed operators, see Kato [1], Lemmas 324, 335, Goldberg [1], Theorem 5.1, and Rota [1], Theorem 1.1. Browder [2], Theorem 1.2, gave similar results for closed linear operators in Fréchet spaces.

II.4.1 Lemma. *If T' has a bounded inverse (T not necessarily closed) and S is the* 1*-ball in X, then $\overline{TS} \supset S_Y{}^0(r)$, where $r = 1/\|(T')^{-1}\|$ and $S_Y{}^0(r) = \{y \mid \|y\| < r\}$.*

Proof. Suppose $y \in S_Y{}^0(r)$ but $y \notin \overline{TS}$. Since $\overline{TS}$ is closed and convex, there exists, by Theorem I.5.10, a nonzero $y' \in Y'$ such that $\operatorname{Re} y'(y) \geq \operatorname{Re} y'(\overline{TS})$. Assert that $\operatorname{Re} y'(y) \geq |y'Tx|$ for all $x \in S \cap \mathfrak{D}(T) = S_1$. Indeed, if $x \in S_1$ and $y'Tx$ is written in polar form $|y'Tx|e^{i\theta}$, then $e^{-i\theta}x \in S_1$. Hence

$$\operatorname{Re} y'(y) \geq \operatorname{Re} y'T(e^{-i\theta}x) = |y'Tx|$$

Consequently, $y' \in \mathfrak{D}(T')$ and

$$\|y'\|\,\|y\| \geq |y'y| \geq \sup_{x \in S_1} |y'Tx| = \|T'y'\| \geq r\|y'\|$$

Thus $\|y\| \geq r$, which contradicts the supposition that $y \in S_Y{}^0(r)$.

II.4.2 Lemma. *Suppose $TS \supset S_Y{}^0(r)$, where S is the* 1*-ball in X and $r > 0$. If T^{-1} exists, then it is continuous with $\|T^{-1}\| \leq 1/r$.*

Proof. If T^{-1} exists, then for $x \neq 0$ and $0 < \varepsilon < 1$, $(1 - \varepsilon)rTx/\|Tx\|$ is in $S_Y{}^0(r)$. By hypothesis, there exists some $z \in S \cap \mathfrak{D}(T)$ such that $Tz = (1 - \varepsilon)rTx/\|Tx\|$. Since T is 1-1, $z = (1 - \varepsilon)rx/\|Tx\|$. Hence $\|Tx\| \geq (1 - \varepsilon)r\|x\|$, which implies that T^{-1} is continuous with $\|T^{-1}\| \leq 1/r$.

II.4.3 Theorem. *Suppose X is complete. If T is closed and T' has a bounded inverse, then $TS \supset S_Y{}^0(r)$, where $r = 1/\|(T')^{-1}\|$ and S is the* 1*-ball of X. Thus $\mathfrak{R}(T) = Y$ and T is an open map. If T^{-1} exists, it is continuous and $S_Y{}^0(1/\|T^{-1}\|)$ is the largest open a-ball which is contained in TS.*

Proof. All but the last statement of the theorem are immediate consequences of Lemmas II.4.1, II.1.7, and II.4.2. Suppose T^{-1} exists and $S_Y{}^0(a) \subset TS$. Then from Lemma II.4.2 and Theorems II.3.9 and

II.2.8, it follows that

$$a \leq \frac{1}{\|T^{-1}\|} = \frac{1}{\|(T')^{-1}\|} = r$$

As a summary of Theorems II.3.11, II.3.13, and II.4.3, we have the following dual results. Note that neither X nor Y need be complete, nor do we require that T be closed in (*i*).

II.4.4 Theorem

i. $\mathcal{R}(T') = X'$ *if and only if* T *has a bounded inverse.*
ii. Suppose that X *and* Y *are complete and that* T *is closed. Then* $\mathcal{R}(T) = Y$ *if and only if* T' *has a bounded inverse.*

II.4.5 Corollary (Banach-Mazur). *If* Y *is a separable Banach space, then there exists a continuous linear operator mapping* l_1 *onto* Y. *Moreover, the conjugate of* Y *is equivalent to a subspace of* l_∞.

Proof. Let $\{y_k\}$ be a sequence of elements of norm 1 which is dense in the unit sphere of Y. Define $T \in [l_1, Y]$ by $T(\{\alpha_k\}) = \sum_{k=1}^{\infty} \alpha_k y_k$. Let $u_1 = (1, 0, 0, \ldots)$, $u_2 = (0, 1, 0, 0, \ldots)$, etc. Then for $y' \in Y'$

$$\|T'y'\| = \sup_k |T'y'(u_k)| = \sup_k |y'y_k| = \sup_{\|y\|=1} |y'y| = \|y'\|$$

Hence T' is an isometry and the range of T is Y by Theorem II.4.4.

II.4.6 Definition. *The 1-1 operator* $\hat{T}$ *induced by* T *is the operator from* $\mathfrak{D}(T)/\mathfrak{N}(T)$ *into* Y *defined by*

$$\hat{T}[x] = Tx$$

Note that $\hat{T}$ is 1-1 and linear with the same range as T.

The importance of considering $\hat{T}$ is that certain results which hold for 1-1 linear operators may be applied to $\hat{T}$ in order to obtain information about T. The proof of Corollary II.4.8 is a case in point.

The next lemma shows that $\hat{T}$ has some of the essential properties of T.

II.4.7 Lemma. *Suppose* $\mathfrak{N}(T)$ *is closed and* $\hat{T}$ *is the 1-1 operator induced by* T. *Then*

i. T *is closed if and only if* $\hat{T}$ *is closed.*
ii. T *is continuous if and only if* $\hat{T}$ *is continuous, in which case* $\|T\| = \|\hat{T}\|$.
iii. $\mathfrak{D}((\hat{T})') = \mathfrak{D}(T')$ *and* $\|(\hat{T})'y'\| = \|T'y'\|$

Proof of (i). Let T be closed. Suppose $[x_n] \to [x]$, $[x_n] \in \mathfrak{D}(T)/\mathfrak{N}(T)$, and $\hat{T}[x_n] \to y$. Then there exists a sequence $v_n \in \mathfrak{N}(T)$ such that $x_n - v_n \to x$. Since $T(x_n - v_n) = \hat{T}[x_n] \to y$ and T is closed, x is in $\mathfrak{D}(T)$ and $Tx = y$. Thus $[x] \in \mathfrak{D}(\hat{T})$ and $\hat{T}[x] = y$, which proves that $\hat{T}$ is closed. Conversely, let $\hat{T}$ be closed. Suppose $x_n \to x$ and $Tx_n \to y$. Then $[x_n] \to [x]$ and $\hat{T}[x_n] = Tx_n \to y$. Hence $[x] \in \mathfrak{D}(\hat{T})$ and $\hat{T}[x] = y$, or, equivalently, $x \in \mathfrak{D}(T)$ and $Tx = y$. Therefore T is closed.

Proof of (ii). Suppose T is continuous. Then

$$\|\hat{T}[x]\| = \|Tz\| \leq \|T\| \, \|z\| \qquad z \in [x]$$

Hence

$$\|\hat{T}[x]\| \leq \|T\| \inf_{z \in [x]} \|z\| = \|T\| \, \|[x]\|$$

Thus $\|\hat{T}\| \leq \|T\|$. On the other hand, if $\hat{T}$ is continuous, then

$$\|Tx\| = \|\hat{T}[x]\| \leq \|\hat{T}\| \, \|[x]\| \leq \|\hat{T}\| \, \|x\|$$

Therefore $\|\hat{T}\| \leq \|T\|$. Combining these results, we obtain $\|T\| = \|\hat{T}\|$.

The proof of *(iii)* follows easily upon replacing T by $y'T$ and $\hat{T}$ by $y'\hat{T}$ in the proof of *(ii)*.

***II.4.8* Corollary.** *Let Y be complete and let T be closed. If T' is continuous, then $\mathfrak{D}(T) = X$.*

If both X and Y are complete and T is closed, then the following three statements are equivalent.

i. *T' is continuous.*
ii. *T is continuous on all of X.*
iii. *$\mathfrak{D}(T')$ is closed.*

Proof. Suppose T' is continuous. Assume, first, that T is 1-1. Let $Y_1 = \overline{\mathfrak{R}(T)}$ and let T_1 be the operator T mapping $\mathfrak{D}(T)$ into Y_1. Then, by Theorems II.3.7 and II.3.9, $(T_1^{-1})'$ has an inverse and

$$T_1' = [(T_1^{-1})^{-1}]' = [(T_1^{-1})']^{-1} \tag{1}$$

Since T' is continuous, it follows that T_1' is also continuous. Thus (1) shows that $(T_1^{-1})'$ has a bounded inverse. Since T is closed, it follows that T_1^{-1} is also closed. Thus we may apply Theorem II.4.3 to T_1^{-1} and obtain $X = \mathfrak{R}(T_1^{-1}) = \mathfrak{D}(T)$. If T is not 1-1, then the 1-1 operator $\hat{T}$ induced by T is closed and its conjugate is bounded by Lemma II.4.7. Thus, by what has just been shown, $X/\mathfrak{N}(T) = \mathfrak{D}(\hat{T}) = \mathfrak{D}(T)/\mathfrak{N}(T)$. Hence $X = \mathfrak{D}(T)$.

If, in addition, X is complete, then T is continuous by the closed-graph theorem. Hence $\mathfrak{D}(T') = Y'$. This shows that *(i)* implies *(ii)* and *(ii)* implies *(iii)*. Assuming *(iii)*, *(i)* is a consequence of the closed-graph theorem applied to T'.

II.4.9 Corollary. *Suppose X is reflexive. If there exists a bounded linear operator which maps X onto Banach space Y, then Y is reflexive.*

If M is a closed subspace of X, then X/M is reflexive.

Proof. Assume that T is an isomorphism from X onto Y. It follows from Theorem II.4.4 that T'' is also an isomorphism. (One can also easily prove directly that T'' is an isomorphism.) Since $T''J_X = J_YT$ and both J_X and T'' are surjective, it follows that

$$Y'' = \mathcal{R}(T''J_X) = \mathcal{R}(J_YT) \subset \mathcal{R}(J_Y)$$

Thus Y is reflexive. Let us only assume that $\mathcal{R}(T) = Y$. Then by Theorem II.3.13, $T'Y'$ is isomorphic to Y'. Since Y' is complete, $T'Y'$ is also complete and therefore a closed subspace of reflexive space X'. Hence $T'Y'$ is reflexive by Theorem I.6.12. Thus, by what was just proved, Y' is reflexive and therefore so is Y. The last statement of the theorem now follows, since the map from X onto X/M defined by $x \to [x]$ is bounded and linear.

One of the most important consequences of Theorem II.4.4 is Theorem IV.1.2, which is proved when closed operators with closed range are discussed.

We come to the final theorem which is needed to complete the state diagram for closed operators.

II.4.10 Theorem. Suppose X is reflexive. If T is closed and 1-1, then $\mathcal{R}(T')$ is dense in X'.

Proof. Let T_1 be the operator T considered as a map onto

$$Y_1 = \mathcal{R}(T)$$

It is easy to see that $\mathcal{R}(T') = \mathcal{R}(T_1')$. By Theorems II.3.7 and II.3.9, T_1' has an inverse and $(T_1')^{-1} = (T_1^{-1})'$. Since T is closed, it is clear that T_1 is closed and therefore so is T_1^{-1}. Applying Theorem II.2.14 to T_1^{-1}, we have that $\mathfrak{D}((T_1^{-1})')$ is dense in X'. Since

$$\mathfrak{D}((T_1^{-1})') = \mathfrak{D}((T_1')^{-1}) = \mathcal{R}(T_1') = \mathcal{R}(T')$$

the theorem follows.

At times it will be convenient for us, throughout the remainder of the book, to refer to the two state diagrams rather than to the various theorems in Secs. II.3 and II.4. For example, an inspection of Diagram II.3.14 shows that $T \in \mathrm{III}_1$ if and only if $T' \in \mathrm{I}_3$. If T is closed, X reflexive, and Y

complete, then an inspection of II.4.11 shows that $T \in \mathrm{II}_3$ if and only if $T' \in \mathrm{III}_2$.

II.4.11 State diagram for closed linear operators

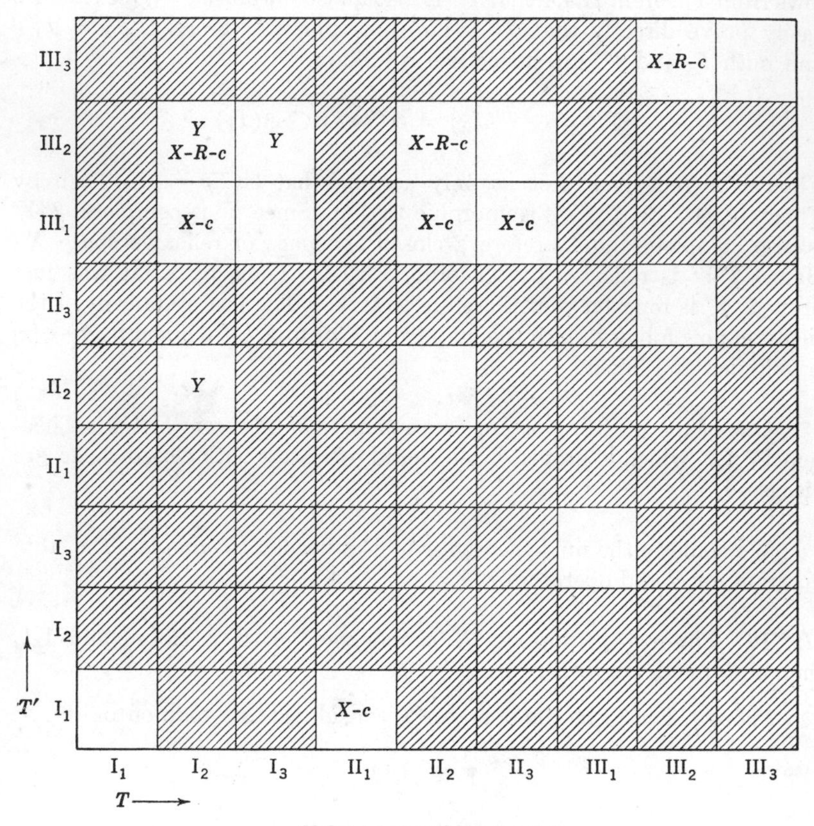

Y: Cannot occur if Y is complete.

X-c: Cannot occur if X is complete and T is closed.

X-R-c: Cannot occur if X is reflexive and T is closed.

II.5 EXAMPLES OF STATES

In this section examples are given which exhibit those states of (T, T') which can occur even when both X and Y are reflexive. Examples are also given which show that the states of (T, T') corresponding to the squares with entries can occur when the corresponding hypotheses on X or Y are removed.

Since some of the arguments depend on properties of compact operators, it is suggested that the reader who is unfamiliar with the theory of compact operators return to this section after Chap. III.

Where possible, compact linear operators are constructed.

$$X = Y = l_2 \qquad T \text{ continuous on } X$$

(I_1, I_1): Let T be the identity operator on X.

(I_3, III_1): Let T be the left-shift operator defined by $T(\{x_k\}) = \{x_{k+1}\}$. Obviously, $T \in \mathrm{I}_3$ and by II.3.14, $T' \in \mathrm{III}_1$.

(III_1, I_3): Let T be the right-shift operator defined by

$$T(\{x_k\}) = \{x_{k-1}\} \qquad 1 < k$$

where $x_0 = 0$. $T \in \mathrm{III}_1$ and by II.4.11, $T' \in \mathrm{I}_3$.

It is easy to see that the left-shift and right-shift operators are conjugates of each other.

$$X = Y = l_2 \qquad T \text{ compact on } X$$

(II_2, II_2): Let T be defined by $T(\{x_k\}) = \{x_k/k\}$. T is compact since $T_n \to T$, $n = 1, 2, \ldots$, where T_n is the compact operator from X into Y defined by $T_n(\{x_k\}) = \{y_k\}$, $y_k = x_k/k$, $1 \leq k \leq n$, and $y_k = 0$, $k > n$. T_n is compact since it is bounded and its range is finite-dimensional. Clearly, $\mathcal{R}(T)$ is dense in Y, and T has an unbounded inverse. Since a compact operator cannot have an infinite-dimensional range which is complete, $T \in \mathrm{II}_2$. From II.4.11, $T' \in \mathrm{II}_2$.

(II_3, III_2): Let L be the left-shift operator on l_2, and let A be the compact operator in example (II_2, II_2). Define $T = AL$. Then T is compact and in II_3. From II.4.11, $T' \in \mathrm{III}_2$.

(III_2, II_3): Let T be the conjugate of the operator in example (II_3, III_2). Since the conjugate of a compact operator is compact, T is compact and from II.4.11, $T \in \mathrm{III}_2$ and $T' \in \mathrm{II}_3$.

(III_3, III_3): Let T be the zero operator. Less trivially, let T_1 be the compact operator in example (II_2, II_2) and let L and R be the left-shift and right-shift operators, respectively. Then $T = T_1RL$ is a compact operator in III_3. By II.4.11, $T' \in \mathrm{III}_3$.

$$X = l_2 \qquad Y \text{ not complete} \qquad T \text{ compact on } X$$

The next example depends on the following observation.

Suppose T is a compact linear operator from reflexive space X into Y. Let T_1 be the operator T considered as a map from X onto $Y_1 = \mathcal{R}(T)$.

Then T_1 is also compact. To see this, suppose $\{x_n\}$ is a bounded sequence in X. Since T is compact, there exists a subsequence $\{v_n\}$ of $\{x_n\}$ such that $Tv_n \to y \in Y$. By Theorem I.6.15, there exists a subsequence $\{z_n\}$ of $\{v_n\}$ which converges weakly to some $x \in X$. The continuity of T implies that $Tz_n \overset{w}{\to} Tx$. Since $Tz_n \to y$, it is clear that $Tz_n \overset{w}{\to} y$. Hence $y = Tx$, since $y'y = y'Tx$ for all $y' \in Y'$. Thus $T_1v_n \to Tx \in Y_1$, showing that T_1 is compact.

(I_2, II_2): Let T_0 be the compact operator in example (II_2, II_2) and let T be the operator T_0 considered as a map from X onto $R(T_0)$. Then T is a compact operator in I_2, and by II.4.11, $T' \in II_2$.

(I_3, III_2): Let L be the left-shift operator on l_2. Choose K as the operator in the preceding example (I_2, II_2). Defining $T = KL$, T is a compact operator in I_3. Hence T' is also compact. Since a compact operator on an infinite-dimensional normed linear space cannot have a bounded inverse, $T' \notin 1$. Thus $T \in III_2$, by II.4.11.

X not complete $\qquad$ $Y = l_2$ $\qquad$ *T continuous on X*

(I_2, III_1): Let X be the normed linear space obtained by renorming l_2 as in Example II.1.10. Define T as the identity map from X onto l_2. Then $T \notin 1$; otherwise the incomplete space X would be isomorphic to Banach space Y. Thus $T \in I_2$ and $T' \in III_1$ by II.4.11.

(II_1, I_1): Choose X as any proper subspace dense in $Y = l_2$ with T as the identity map on X.

(II_2, III_1): Let A be the operator in example (I_2, III_1). Setting X_1 as the domain of A and Y_1 as any proper subspace dense in l_2, define $T: X_1 \times Y_1 \to l_2 \times l_2$ by $T(x, y) = (Ax, y)$. Take the norm on $l_2 \times l_2$ to be given by $\|(x, y)\| = (\|x\|^2 + \|y\|^2)^{\frac{1}{2}}$. With respect to the inner product

$$\langle (x, y), (u, v) \rangle = \langle x, u \rangle + \langle y, v \rangle$$

$l_2 \times l_2$ is a separable Hilbert space which, by Theorem I.7.24, is equivalent to l_2. T is easily seen to be in II_2. Given $z' \in (l_2 \times l_2)$, define x' and $y' \in l_2'$ by $x'(x) = z'(x, 0)$ and $y'(y) = z'(0, y)$. Then

$$|z'(x, y)| \leq \|x'\| \, \|x\| + \|y'\| \, \|y\| \leq (\|x'\| + \|y'\|) \|(x, y)\|$$

Hence $\|z'\| \leq \|x'\| + \|y'\|$. For $\|x\| = 1$,

$$\|T'z'\| \geq |T'z'(x, 0)| = |(A'x')x|$$

Thus

$$\|T'z'\| \geq \|A'x'\| \geq \frac{\|x'\|}{\|(A')^{-1}\|} \tag{1}$$

Similarly,

$$\|T'z'\| \geq |T'z'(0, y)| = |y'y| \qquad \|y\| = 1$$

Thus

(2) $$\|T'z'\| \geq \|y'\|$$

Hence, by (1) and (2), there exists some $m > 0$ such that

$$\|T'z'\| \geq m(\|x'\| + \|y'\|) \geq m\|z'\|$$

This shows $T' \in 1$. It follows from II.4.11 that $T' \in \mathrm{III}_1$.

(II_3, III_1): Let A, X_1, and Y_1 be as above. Define $T: X_1 \times Y_1 \rightarrow l_2 \times l_2$ by $T(x, y) = (Ax, Ly)$, where L is the left-shift operator on l_2. T is easily seen to be in II_3. Arguing as in the above example, we obtain

$$\|T'z'\| \geq \|L'y'\| = \|y'\|$$

Recall that L' is the right-shift operator and, in particular, is an isometry. Thus $T' \in \mathrm{III}_1$, by II.4.11.

X complete but not reflexive $\qquad Y = l_2 \qquad$ *T compact on X*

(II_2, III_2): If $0 < p \leq q \leq \infty$, then $x = (x_1, x_2, \ldots) \in l_p$ implies $x \in l_q$ and $\|x\|_p \geq \|x\|_q$. Indeed, if $q < \infty$, then $\|x\|_q^q = \sum_{k=1}^{\infty} |x_k|^p |x_k|^{q-p} < \infty$, since $\{x_k\}$ is bounded. Now $\|x\|_q \geq |x_k|, 1 \leq k$. Thus $\|x\|_q^q \leq \|x\|_q^{q-p}\|x\|_p^p$ whence $\|x\|_q \leq \|x\|_p$. If $q = \infty$, then clearly $\|x\|_\infty \leq \|x\|_p$. Let A be the identity map from l_1 into l_2. Then by what was just proved, A is continuous. Define $T: l_1 \rightarrow l_2$ by $T = KA$, where K is the compact operator in example (II_2, II_2). Since T' is compact, it is not in 1 and its range is separable. However, l_∞ is not separable. Thus $T' \in \mathrm{III}_2$ by II.4.11.

(III_2, III_3): Let T_1 be the compact operator T in the preceding example and let R be the right-shift operator on l_1. Define $T: l_1 \rightarrow l_2$ as the operator $T_1 R$. Then T is clearly in III_2. Since compact operator T' has a separable range, T' is in III. Thus $T' \in \mathrm{III}_3$, by II.4.11.

The next example is for the only square in II.4.11 which has not been accounted for.

X complete but not reflexive $\qquad$ *Y not complete* $\qquad$ *T compact on X*

(I_2, III_2): Let T_1 be the operator from $l_1 \rightarrow l_2$ defined by

$$T_1(\{x_k\}) = \{x_k/k\}$$

Then by the argument given in example (II_2, II_2), T_1 is a compact operator with an unbounded inverse. Define T to be the operator T_1 considered as a map from l_1 onto $Y = \mathfrak{R}(T_1)$. Then $T \in I_2$. We show that T is compact. Suppose $\{x_n\}$ is a sequence of elements in the 1-sphere of l_1. Since T_1 is compact, there exists a subsequence $\{v_n\}$ of $\{x_n\}$ such that $Tv_n \to y = (y_1, y_2, \ldots) \in \mathfrak{R}(T_1)$. Writing $v_n = (v_1{}^n, v_2{}^n, \ldots)$, it follows that $\lim_{n\to\infty} v_k{}^n = ky_k$, $1 \le k$. Hence, for any positive integer M,

$$\sum_{k=1}^{M} |ky_k| \le \sum_{k=1}^{M} |ky_k - v_k{}^n| + \sum_{k=1}^{M} |v_k{}^n|$$

$$\le \sum_{k=1}^{M} |ky_k - v_k{}^n| + 1 \to 1 \qquad \text{as } n \to \infty$$

Thus $\sum_{k=1}^{M} |ky_k| \le 1$ for all M, and therefore $x = \{ky_k\}$ is an element in l_1. Hence $Tv_n \to y = Tx$, which shows that T is compact. Since T' is compact, its range is a separable subspace of inseparable space l_∞. Thus $T' \in III$. It follows from II.4.11 that T' is in III_2.

Since the operators in the above examples are all continuous on X, we have also shown that the state diagram for closed linear operators is the same as the state diagram for continuous linear operators defined on all of X. In order to gain some insight into the reason for this, we mention the following theorem which was proved by Goldberg in [2].

Roughly speaking, it is shown that by letting E be $\mathfrak{D}(T')$ with norm defined by $\|y'\| + \|T'y'\|$, E becomes a conjugate space. Moreover, T_1', which is the bounded operator T' with respect to E, becomes the conjugate of a bounded operator T_1. Finally, T_1 and T_1' have the same states as T and T', respectively. Thus the state diagram for (T, T') is the same as the state diagram for (T_1, T_1').

II.5.1 Theorem. *Suppose T is closed. Let E denote the space $\mathfrak{D}(T')$ with norm $\|y'\|_1 = \|y'\| + \|T'y'\|$ and let T'_E be the operator T' mapping E into X'. Define $J: Y \to E'$ by $(Jy)y' = y'y$ and $A: E \to (JY)'$ by $(Ay')Jy = (Jy)y'$. Then*

i. *E is linearly isometric to $(JY)'$ under the map A.*
ii. *JT is a continuous linear map from $\mathfrak{D}(T)$ into JY.*
iii. *$T'_E = (JT)'A$.*
iv. *T and T' are in the same states as bounded operators JT and $(JT)'$, respectively.*

Suppose that X is reflexive and Y is complete. Let Z be the closure of JY in E' and let $\overline{JT}: X \to Z$ be the continuous linear extension of JT to all of X. Then T and T' are in the same states as $\overline{JT}$ and $(\overline{JT})'$, respectively.

A state diagram for linear operators in a topological linear space setting was constructed by Krishnamurthy [1].

Six examples which demonstrate the possibility of states corresponding to the blank squares still unaccounted for in Diagram II.3.14 may be found in Goldberg [1].

II.5.2 Definition. *Let $X = Y$ and let I be the identity operator on X. A scalar λ belongs to*

i. $\rho(T)$, *called the* **resolvent** *of T, if $T - \lambda I$ has a dense range and a bounded inverse.*
ii. $\sigma(T)$, *called the* **spectrum** *of T, if $\lambda \notin \rho(T)$.*

The set $\sigma(T)$ is decomposed into the following disjoint sets by letting λ belong to

iii. $P\sigma(T)$, *called the* **point spectrum** *of T, if $T - \lambda I \in 3$.*
iv. $R\sigma(T)$, *called the* **residual spectrum** *of T, if $T - \lambda I \in$ III but not in 3.*
v. $C\sigma(T)$, *called the* **continuous spectrum** *of T, if $T - \lambda I \in 2$ but not in III.*

Since $T' - \lambda I = (T - \lambda I)'$ and $T - \lambda I$ is closed when T is closed, an appeal to the state diagrams for $(T - \lambda I, (T - \lambda I)')$ verifies the following relations.

II.5.3 Corollary

i. $\sigma(T) = \sigma(T')$
ii. $P\sigma(T) \subset R\sigma(T') \cup P\sigma(T')$
iii. $P\sigma(T') \subset R\sigma(T) \cup P\sigma(T)$
iv. $C\sigma(T) \subset R\sigma(T') \cup C\sigma(T')$
v. $C\sigma(T') \subset C\sigma(T)$
vi. $R\sigma(T) \subset P\sigma(T')$
vii. $R\sigma(T') \subset C\sigma(T) \cup P\sigma(T)$
viii. *If T is closed and X is reflexive, then $C\sigma(T) = C\sigma(T')$ and $R\sigma(T') \subset P\sigma(T)$*

II.5.4 Remark. The same argument used to prove the above corollary can be used to prove the following generalization.

Let B be a continuous linear operator with $X \supset \mathfrak{D}(B) \supset \mathfrak{D}(T)$ and range in Y. (Y not necessarily X.) Then $(T - \lambda B)' = T' - \lambda B'$. Define $\rho_B(T)$ as in (*i*) of Definition II.5.2, with B replacing I. Similarly, define $\sigma_B(T)$, etc. Then (*i*) through (*vii*) of Corollary II.5.3 hold, where the symbol $\sigma_{B'}(T')$ replaces $\sigma(T')$ wherever it appears; for example, (*ii*) now reads $P\sigma_B(T) \subset R\sigma_{B'}(T') \cup P\sigma_{B'}(T')$.

II.6 CHARACTERIZATIONS OF STATES OF OPERATORS

Motivated by certain important concepts in the study of Banach algebras, we show that the states of closed linear operators can be characterized in terms of these concepts.

II.6.1 Definition

i. *T is called* **left (right) regular** *provided there exists an operator $A \in [Y, X]$ such that $AT = I$, $(TA = I)$, where I is the identity map on $\mathfrak{D}(T)$ (Y). The set of all left (right) regular elements will be denoted by R_l (R_r).*

ii. *T is called a* **left (right) divisor of zero** *if there exists an $A \neq 0$ in $[Y, X]$ such that $TA = 0$ $(AT = 0)$ on Y $(\mathfrak{D}(T))$. The set of all left (right) divisors of zero is denoted by D_l (D_r).*

iii. *T is called a* **left (right) topological divisor of zero** *if there exists a sequence $\{A_n\}$ in $[Y, X]$ such that $\|A_n\| = 1$ and $TA_n \to 0$ $(A_nT \to 0)$ in $[Y]$ $([\mathfrak{D}(T), X])$. The set of all left (right) topological divisors of zero is denoted by Z_l (Z_r).*

For convenience, we shall use notations such as $\text{III} = D_r$ to mean that T is in state III if and only if $T \in D_r$ and $1 = \complement Z_l$ to mean that $T \in 1$ if and only if T is in the complement of the set Z_l; that is, if T is not a left topological divisor of zero.

II.6.2 Theorem

i. $\text{III} = D_r$

ii. $1 = \complement Z_l$

iii. $3 = D_l$

iv. $2 = Z_l \cap \complement D_l$

Proof of (i). If $T \in \text{III}$, there exists some $y' \in Y'$ such that $y' \neq 0$ and $y'\mathfrak{R}(T) = 0$. Define $A \in [Y, X]$ by $Ay = y'(y)x_0$, where $x_0 \neq 0$ is fixed in $\mathfrak{D}(T)$. Then $A \neq 0$ and $ATx = 0$ for all $x \in \mathfrak{D}(T)$. Thus $T \in D_r$. Conversely, suppose $T \in D_r$. Let $A \neq 0$ be in $[Y, X]$ and let $AT = 0$. Then $\mathfrak{R}(T) \subset \mathfrak{N}(A) \neq Y$, whence $T \in \text{III}$.

Proof of (ii). Suppose $T \in 1$ and suppose there exists a sequence $\{A_n\}$ in $[Y, X]$ such that $A_n \neq 0$ and $TA_n \to 0$ in $[Y]$. Now there exists an $m > 0$ such that for each y in the 1-sphere of Y,

$$\|TA_n\| \geq \|TA_ny\| \geq m\|A_ny\|$$

Hence $A_n \to 0$ in $[Y, X]$ and therefore $T \notin Z_l$. On the other hand, suppose $T \notin 1$. Then there exists a sequence $\{x_n\}$ in $\mathfrak{D}(T)$ such that $\|x_n\| = 1$ and $Tx_n \to 0$. Fix $y' \in Y'$ with $\|y'\| = 1$. For each n, define $A_n \in [Y, X]$

by $A_n y = y'(y)x_n$. Then $\|A_n\| = 1$ and $\|TA_n\| \leq \|Tx_n\| \to 0$. Hence $T \in Z_l$.

Proof of (iii). Suppose there exists and $x \neq 0$ such that $Tx = 0$. Define $A \in [Y, X]$ by $Ay = y'(y)x$, where y' is some nonzero element in Y'. Then $A \neq 0$ but $TA = 0$. Hence $T \in D_l$. If $T \in D_l$, there exists an $A \neq 0 \in [Y, X]$ such that $TA = 0$. Consequently, $(0) \neq \mathcal{R}(A) \subset \mathfrak{N}(T)$.

(iv) is a trivial consequence of *(ii)* and *(iii)*.

II.6.3 Theorem. *Suppose X and Y are complete and T is closed.* Then I $= \mathcal{C}Z_r$.

Proof. Suppose $T \in$ I and also in Z_r. Then there exist sequences $\{A_n\}$ in $[Y, X]$ and $\{y_n\} \in Y$ such that

$$\|A_n\| = 1 \qquad \|y_n\| = 1 \qquad \|A_n y_n\| \geq 1 - \frac{1}{n}$$

$$\|A_n T\| \to 0 \qquad \text{in } [\mathfrak{D}(T), X]$$

Let $\hat{T}$ be the 1-1 operator induced by T. The proof that $\|A_n \hat{T}\| = \|A_n T\|$ is the same as the proof that $\|\hat{T}\| = \|T\|$ in Lemma II.4.7. Moreover, $\hat{T}$ is closed. Since $T \in$ I, there exists a sequence $\{x_n\}$ such that $Tx_n = y_n$. The sequence $\{[x_n]\}$ in $\mathfrak{D}(T)/\mathfrak{N}(T)$ is unbounded since $\|A_n T\| \to 0$ and

$$1 - \frac{1}{n} \leq \|A_n y_n\| = \|A_n \hat{T}[x_n]\| \leq \|A_n T\| \, \|[x_n]\|$$

Thus, for $z_n = [x_n]/\|[x_n]\|$,

$$\|\hat{T} z_n\| = \frac{\|Tx_n\|}{\|[x_n]\|} = \frac{\|y_n\|}{\|[x_n]\|} = \frac{1}{\|[x_n]\|} \to 0$$

This implies that $\hat{T}$ does not have a bounded inverse, which contradicts Theorem II.1.9. Hence $T \in$ I implies $T \in \mathcal{C}Z_r$.

Conversely, suppose $T \notin Z_r$. To show $T \in$ I, it suffices to show that $T' \in$ 1 by II.4.11. Assume T' does not have a bounded inverse. Then there exists a sequence $\{y'_n\} \in \mathfrak{D}(T')$ such that $\|y'_n\| = 1$ and $T'y'_n \to 0$. Let x_0 be an element of norm 1 in $\mathfrak{D}(T)$. Define $A_n \in [Y, X]$ by

$$A_n y = y'_n(y) x_0$$

Then $\|A_n\| = 1$, and for all x in the 1-sphere of $\mathfrak{D}(T)$,

$$\|A_n T x\| \leq \|T' y'_n\| \to 0$$

Hence $\|A_n T\| \to 0$, which contradicts the supposition $T \notin Z_r$.

II.6.4 Theorem. *If X and Y are complete and T is closed, then the following statements are equivalent.*

i. $T \in \mathrm{I}_1$
ii. $T \notin Z_l \cup Z_r$
iii. $T \in R_l \cap R_r$

Proof. (*i*) is equivalent to (*ii*) by Theorems II.6.2 and II.6.3. We show that (*i*) is equivalent to (*iii*). If $T \in \mathrm{I}_1$, then $T^{-1} \in [Y, X]$, $TT^{-1} = I$ on Y, and $T^{-1}T = I$ on $\mathfrak{D}(T)$. Thus, by definition, $T \in R_l \cap R_r$. If $T \in R_l \cap R_r$, then clearly T is 1-1 and $\mathfrak{R}(T) = Y$. Hence from II.4.11, $T \in \mathrm{I}_1$.

Combining Theorems II.6.2 to II.6.4 we obtain the following result.

II.6.5 Theorem. *Suppose X and Y are complete. Then the states for a closed linear operator with domain dense in X and range contained in Y are characterized by*

i. $\mathrm{I}_1 = R_l \cap R_r = \mathcal{C}(Z_r \cup Z_l)$
ii. I_2 *cannot exist*
iii. $\mathrm{I}_3 = D_l \cap \mathcal{C}Z_r$
iv. II_1 *cannot exist*
v. $\mathrm{II}_2 = Z_r \cap Z_l \cap \mathcal{C}(D_r \cup D_l)$
vi. $\mathrm{II}_3 = Z_r \cap D_l \cap \mathcal{C}D_r$
vii. $\mathrm{III}_1 = D_r \cap \mathcal{C}Z_l$
viii. $\mathrm{III}_2 = D_r \cap Z_l \cap \mathcal{C}D_l$
ix. $\mathrm{III}_3 = D_r \cap D_l$

II.7 ADJOINT OF AN OPERATOR IN HILBERT SPACE

The Riesz representation theorem shows how a Hilbert space can be identified with its conjugate. Consequently, if X and Y are Hilbert spaces, then T' can be identified with an operator mapping a subspace of Y into X as follows.

II.7.1 Definition. *Let X and Y be Hilbert spaces. Let E_X and E_Y be the isometries from X onto X' and Y onto Y', respectively, as in Definition* I.7.19. *The **adjoint** of T, written T^*, is defined by $T^* = E_X{}^{-1}T'E_Y$.*

II.7.2 Remarks

i. The usual definition of T^*, which is equivalent to the one above, is: $\mathfrak{D}(T^*)$ is the subspace of Y consisting of those y for which $f_y(x) = \langle Tx, y\rangle$ is continuous on $\mathfrak{D}(T)$. For such y let $T^*y = z$, where $f_y(x) = \langle x, z\rangle$ for all $x \in \mathfrak{D}(T)$. Therefore the operator

T^* is characterized by the equation

$$\langle Tx, y \rangle = \langle x, z \rangle = \langle x, T^*y \rangle$$

ii. From the definitions of E_X and E_Y, it is clear that T^* is T' when the spaces are identified with their conjugates. In particular, the state diagrams for (T, T') and (T, T^*) are the same.

iii. If $X = Y$ and $T = T^*$, then T is called *self-adjoint.* Thus, a self-adjoint operator is closed, and Diagram II.4.11 shows that I_1, II_2 and III_3 are its only possible states. For a treatment of self-adjoint ordinary differential operators, the reader is referred to Neumark [1] and Dunford and Schwartz [1], part II.

chapter III

STRICTLY SINGULAR OPERATORS

The concept of a strictly singular operator was introduced by Kato [1] in his treatment of perturbation theory. It was shown that certain properties of T are shared by operators of the form $T + B$, where B is strictly singular. In this chapter, properties of strictly singular operators and, in particular, compact operators are obtained, the results of which are used in Chap. V.

Throughout this chapter, X and Y denote normed linear spaces. Completeness is assumed only when specifically stated.

III.1 COMPACT OPERATORS

III.1.1 Definition. *Let B be a bounded linear operator with domain in X and range in Y. B is called* **strictly singular** *if it does not have a bounded inverse on any infinite-dimensional subspace contained in its domain.*

If B is defined on all of X, it suffices to consider only closed infinite-dimensional subspaces.

The most important examples of strictly singular operators are the compact operators. These play a major role in the study of differential and integral equations. In cases where the normed linear spaces are not assumed complete, it is convenient to consider also precompact operators.

III.1.2 Definition. *Let K be a linear operator with domain in X and range in Y. If KS is totally bounded in Y, where S is the 1-ball in X, then K is called* **precompact.** *If $\overline{KS}$ is compact in Y, then K is called* **compact.**

III.1.3 Theorem. *Every precompact operator is strictly singular.*

Proof. Let B be a precompact operator with domain in X and range in Y. B is bounded since a totally bounded set is bounded. Suppose B has a bounded inverse on a subspace $M \subset \mathfrak{D}(B)$. Then BS_M is totally bounded, where S_M is the 1-ball in M. Since B has a bounded inverse on M, it follows that S_M is totally bounded in M. Hence M is finite-dimensional by Theorem I.4.6.

Examples of strictly singular operators which are not compact are given in Sec. III.3.

III.1.4 Remarks

i. Every bounded linear operator with finite-dimensional range is compact. For if $K: X \to Y$ is such an operator, then letting S be the 1-ball of X, KS is a bounded set in the finite-dimensional subspace $\mathfrak{R}(K)$ of Y. The compactness of $\overline{KS}$ is therefore a consequence of Corollary I.4.3.

ii. An operator is compact if and only if it takes every bounded sequence into a sequence which has a convergent subsequence. An operator is precompact if and only if it takes every bounded sequence into a sequence which has a Cauchy subsequence.

iii. If K is precompact as a mapping from X into Y and $\hat{Y}$ is the completion of Y, then K, when considered as a map into $\hat{Y}$, is compact by Theorem 0.3. Thus $K \in [X, Y]$ is compact if and only if it is precompact, provided Y is complete.

III.1.5 Lemma. *Suppose $\{K_N\}$ is a sequence of precompact operators in $[X, Y]$. If $K_N \to K$ in $[X, Y]$, then K is precompact.*

Proof. Let $\varepsilon > 0$ be given. By hypothesis, there exists an integer N such that

$$\|K - K_N\| < \frac{\varepsilon}{3} \tag{1}$$

Since K_N is precompact, there exist elements $x_1, x_2, \ldots, x_m$ in the 1-ball S of X such that given $x \in S$, there is an x_i for which

$$\|K_N x - K_N x_i\| < \frac{\varepsilon}{3} \tag{2}$$

Hence, from (1) and (2),

$$\|Kx - Kx_i\| \leq \|Kx - K_Nx\| + \|K_Nx - K_Nx_i\| + \|K_Nx_i - Kx_i\| < \frac{\varepsilon}{3} + \frac{\varepsilon}{3} + \frac{\varepsilon}{3}$$

which implies that K is precompact.

As an application of the lemma, the following example of a compact operator is given.

III.1.6 Example. Let $I = [0, 1]$ and let $1 < p, q < \infty$, with p' and q' conjugate to p and q, respectively. Suppose $k(s, t)$ is in $\mathfrak{L}_r(I \times I)$, where r is the larger of p' and q. Then the linear operator K defined by

$$(Kx)(t) = \int_0^1 k(s, t)x(s)\, ds$$

is compact as a map from $\mathfrak{L}_p(I)$ into $\mathfrak{L}_q(I)$.

Proof. To show, first of all, that K maps $\mathfrak{L}_p(I)$ into $\mathfrak{L}_q(I)$, suppose x is in $\mathfrak{L}_p(I)$. Since $r' \leq p$, we have, by Hölder's inequality, that x is in $\mathfrak{L}_{r'}(I)$, $\|x\|_{r'} \leq \|x\|_p$, and

$$(1) \quad \|Kx\|_q{}^q = \int_0^1 dt \left| \int_0^1 k(s, t)x(s)\, ds \right|^q \leq \int_0^1 dt \left[\left(\int_0^1 |k(s, t)|^r\, ds \right)^{q/r} \|x\|_{r'}{}^q \right]$$

$$\leq \|x\|_p{}^q \int_0^1 dt \left(\int_0^1 |k(s, t)|^r\, ds \right)^{q/r}$$

Since $0 < q/r \leq 1$, Hölder's inequality implies that for any $g \in \mathfrak{L}_1(I)$,

$$(2) \qquad \int_0^1 |g(t)|^{q/r}\, dt \leq \left(\int_0^1 |g(t)|\, dt \right)^{q/r}$$

Taking $g(t) = \int_0^1 |k(s, t)|^r\, ds$, we obtain from (1) and (2) that

$$(3) \qquad \|Kx\|_q \leq \|x\|_p \left(\int_0^1 \int_0^1 |k(s, t)|^r\, ds\, dt \right)^{1/r}$$

Thus K is continuous as a map from $\mathfrak{L}_p(I)$ into $\mathfrak{L}_q(I)$.

The operator K will next be shown to be compact under the additional assumption that k is continuous on $I \times I$. Let $\{x_n\}$ be a bounded sequence in $\mathfrak{L}_p(I)$. Then for $y_n = Kx_n$, we have from the inequality $\|x_n\|_1 \leq \|x_n\|_p$ that

$$|y_n(t)| \leq \int_0^1 |k(s, t)x_n(s)|\, ds \leq \max_{0 \leq s,t \leq 1} |k(s, t)|\, \|x_n\|_p$$

Hence $\{y_n\}$ is uniformly bounded on I. Since k is uniformly continuous on $I \times I$, there exists for each $\varepsilon > 0$ a $\delta = \delta(\varepsilon) > 0$ such that for all $s \in I$,

$$|k(s, t_1) - k(s, t_2)| < \varepsilon \qquad |t_1 - t_2| < \delta$$

For such t_1 and t_2,

$$(4) \quad |y_n(t_1) - y_n(t_2)| \leq \int_0^1 |k(s, t_1) - k(s, t_2)|\, |x_n(s)|\, ds \leq \varepsilon \|x_n\|_p$$

Since $\{x_n\}$ is bounded in $\mathfrak{L}_p(I)$, (4) shows that $\{y_n\}$ is equicontinuous on I. Hence, by the Ascoli-Arzelà theorem, $\{y_n\}$ contains a subsequence converging in $C([0, 1])$ which in turn implies convergence in $\mathfrak{L}_p([0, 1])$. Thus K is compact.

Removing the restriction that k be continuous, we nevertheless know by Theorem 0.9 that there exists a sequence $\{k_n\}$ of functions continuous on $[0, 1] \times [0, 1]$ which converges in $\mathfrak{L}_r(I \times I)$ to k. Let $K_n \in [\mathfrak{L}_p(I), \mathfrak{L}_q(I)]$ be defined by

$$(K_n x)(t) = \int_0^1 k_n(s, t) x(s)\, ds$$

Then by what was just proved, K_n is compact. Substituting $k_n - k$ for k in (3) gives, for each $x \in \mathfrak{L}_p(I)$,

$$\|(K_n - K)x\|_q \leq \|x\|_p \left(\int_0^1 \int_0^1 |k_n(s, t) - k(s, t)|^r\, ds\, dt \right)^{1/r} \to 0 \qquad \text{as } n \to \infty$$

Thus $K_n \to K$ in $[\mathfrak{L}_p(I), \mathfrak{L}_q(I)]$. Consequently, K is compact by Lemma III.1.5.

In general, if Y is not complete, the limit in $[X, Y]$ of compact operators need not be compact. This may be seen in the following example.

III.1.7 Example. Let c_0 be the subspace of l_∞ consisting of the sequences which converge to 0. Let T_0 be the operator mapping c_0 into l_2 defined by

$$T_0(\{\alpha_k\}) = \left\{ \frac{\alpha_k}{k} \right\} \qquad 1 \leq k$$

Define T to be the operator T_0 considered as a map from c_0 onto $Y = \mathfrak{R}(T_0)$. Then T is not compact. For let $x_k = \underbrace{\{1, 1, \ldots, 1}_{k}, 0, 0, \ldots\}$. Then $\|x_k\| = 1$ in c_0, but $Tx_k \to y$ in l_2, where $y = \{1, \frac{1}{2}, \frac{1}{3}, \ldots\}$. Now, y is not in Y, since $Tx = y$ implies $x = \{1, 1, 1, \ldots\}$, which is not in c_0. Hence $\{Tx_k\}$ has no subsequence which converges in Y; that is, T is not compact. However, the sequence of operators $T_n : c_0 \to Y$, defined by

$T_n(\{\alpha_k\}) = \{\beta_k\}$, where $\beta_k = \alpha_k/k$ for $1 \leq k \leq n$ and $\beta_k = 0, k > n$, converges in $[c_0, Y]$ to T. Moreover, $\mathcal{R}(T_n)$ is finite-dimensional and therefore T_n is compact.

Corollary III.1.10 shows how compact operators may arise from closed operators which do not have a closed range.

The following theorem, excluding the assertion that the operator is precompact, is due to Kato [1].

III.1.8 Definition. *A subspace M of a vector space V is said to have **finite deficiency** in V if the dimension of V/M is finite. This is written*

$$\dim \frac{V}{M} < \infty$$

Even though M is not contained in $\mathfrak{D}(T)$, a restriction of T to M will mean a restriction of T to $M \cap \mathfrak{D}(T)$.

III.1.9 Theorem. *Let T be a linear map from a subspace of X into Y. Suppose that T does not have a bounded inverse when restricted to any closed subspace having finite deficiency in X. Then given $\varepsilon > 0$, there exists an infinite-dimensional subspace $M(\varepsilon)$ contained in $\mathfrak{D}(T)$ such that T restricted to $M(\varepsilon)$ is precompact and has norm not exceeding ε.*

Proof. The hypothesis implies the existence of an $x_1 \in X$ such that $\|x_1\| = 1$ and $\|Tx_1\| < 3^{-1}\varepsilon$. There is an $x_1' \in X'$ such that $\|x_1'\| = 1$ and $x_1'x_1 = \|x_1\| = 1$. Since $\mathfrak{N}(x_1')$ has deficiency 1 in X, there exists an $x_2 \in \mathfrak{N}(x_1')$ such that $\|x_2\| = 1$ and $\|Tx_2\| < 3^{-2}\varepsilon$. There exists an $x_2' \in X'$ such that $\|x_2'\| = 1$ and $x_2'x_2 = \|x_2\| = 1$. Since $\mathfrak{N}(x_1') \cap \mathfrak{N}(x_2')$ has finite deficiency in X, there exists an $x_3 \in \mathfrak{N}(x_1') \cap \mathfrak{N}(x_2')$ such that $\|x_3\| = 1$ and $\|Tx_3\| < 3^{-3}\varepsilon$. Inductively, sequences $\{x_k\}$ and $\{x_k'\}$ are constructed in X and X', respectively, with the following properties.

(1) $\quad \|x_k\| = \|x_k'\| = x_k'x_k = 1 \qquad \|Tx_k\| < 3^{-k}\varepsilon, \quad 1 \leq k < \infty$

(2) $\quad x_k \in \bigcap_{i=1}^{k-1} \mathfrak{N}(x_i')$ or, equivalently, $x_i'x_k = 0 \qquad 1 \leq i < k$

It is easy to verify that the set of x_k is linearly independent, whence $M = \text{sp}\,\{x_1, x_2, \ldots\}$ is an infinite-dimensional subspace of $\mathfrak{D}(T)$. It will now be shown that the restriction T_M of T to M has norm not exceeding ε. Suppose $x = \sum_{i=1}^{m} \alpha_i x_i$. Then from (1) and (2),

$$|\alpha_1| = |x_1'x| \leq \|x_1'\|\,\|x\| = \|x\|$$

In fact,

$$|\alpha_k| \leq 2^{k-1}\|x\| \qquad 1 \leq k \leq m \tag{3}$$

For suppose (3) is true for $k \leq j < m$. Then from (1) and (2),

$$x'_{j+1}x = \sum_{i=1}^{j} \alpha_i x'_{j+1}(x_i) + \alpha_{j+1} \tag{4}$$

Hence, by (4) and the induction hypothesis,

$$|\alpha_{j+1}| \leq |x'_{j+1}x| + \sum_{i=1}^{j} |\alpha_i|\, |x'_{j+1}x_i| \leq \|x\| + \sum_{i=1}^{j} 2^{i-1}\|x\| = 2^j\|x\|$$

Thus (3) follows by induction and

$$\|Tx\| \leq \sum_{i=1}^{m} |\alpha_i|\, \|Tx_i\| \leq \sum_{i=1}^{m} 2^{i-1}3^{-i}\varepsilon\|x\| \leq \varepsilon\|x\|$$

Hence $\|T_M\| \leq \varepsilon$. To prove that T_M is precompact, it suffices to show that T_M is the limit in $[M, Y]$ of a sequence of precompact operators. For each positive integer n, define $T_n{}^M : M \to Y$ to be T on sp $\{x_1, x_2, \ldots, x_n\}$ and 0 on sp $\{x_{n+1}, x_{n+2}, \ldots\}$. Clearly $T_n{}^M$ is linear and has finite-dimensional range. Moreover, $T_n{}^M$ is bounded on M; for if $x = \sum_{i=1}^{n+k} \alpha_i x_i$, then by (1) and (3),

$$\|T_n{}^M x\| \leq \sum_{i=1}^{n} |\alpha_i|\, \|Tx_i\| \leq \sum_{i=1}^{n} 2^{i-1}3^{-i}\varepsilon\|x\|$$

Thus $T_n{}^M$ is precompact by III.1.4. Since

$$\begin{aligned}\|T_M x - T_n{}^M x\| &\leq \sum_{i=n+1}^{\infty} |\alpha_i|\, \|Tx_i\| \\ &\leq \varepsilon\|x\| \sum_{i=n+1}^{\infty} 2^{i-1}3^{-i} \to 0 \qquad \text{as } n \to \infty\end{aligned}$$

it follows that $T_n{}^M$ converges to T_M in $[M, Y]$, completing the proof of the theorem.

III.1.10 Corollary. *Let X and Y be complete. If T is closed but $\mathfrak{R}(T)$ is not closed, then for each $\varepsilon > 0$ there exists an infinite-dimensional closed subspace $M(\varepsilon)$ contained in $\mathfrak{D}(T)$ such that T restricted to $M(\varepsilon)$ is compact with norm not exceeding ε.*

Proof. Let W be a closed subspace having finite deficiency in X. Assume T has a bounded inverse on W. Then TW is closed; for if $Tx_n \to y$, $x_n \in W$, then $\{x_n\}$ is a Cauchy sequence and therefore converges to some x in Banach space W. Since T is closed, x is in $\mathfrak{D}(T)$ and $Tx = y$. By hypothesis, there exists a finite-dimensional subspace N of X such that $X = W + N$. Hence $TX = TW + TN$. Since TW is closed and TN is finite-dimensional, TX is also closed by Theorem I.4.12. But this contradicts the hypothesis that $\mathfrak{R}(T)$ is not closed. Therefore T does not have a bounded inverse on W. Hence there exists an $M = M(\varepsilon)$ with the properties described in Theorem III.1.9. Since Y is complete and T is closed and bounded on M, it follows that $\bar{M}$ is contained in $\mathfrak{D}(T)$. Moreover, $\|T_{\bar{M}}\| \leq \varepsilon$ and $\overline{TS_{\bar{M}}} = \overline{TS_M}$, where $S_{\bar{M}}$ and S_M are the 1-balls in $\bar{M}$ and M, respectively, and $T_{\bar{M}}$ is the restriction of T to $\bar{M}$. The precompactess of T and the completeness of Y imply that $\overline{TS_M}$ is compact. Thus $T_{\bar{M}}$ is compact.

It is shown in Example III.3.7 that the conjugate of a strictly singular operator is not necessarily strictly singular and conversely, a bounded operator which has a strictly singular conjugate need not be strictly singular. This lack of duality, fortunately, does not carry over to precompact operators, as the next theorem shows.

III.1.11 Theorem. *A bounded linear operator is precompact if and only if its conjugate is compact.*

Proof. Let K be a precompact linear operator which maps X into Y. We shall prove that K' is precompact and therefore compact since X' is complete. Given $\varepsilon > 0$, there exist $x_1, x_2, \ldots, x_n$ in the 1-ball S_X of X such that for each $x \in S_X$ there is an x_i for which

$$\|Kx - Kx_i\| < \frac{\varepsilon}{3} \tag{1}$$

Let A be the map from Y' to unitary n-space given by

$$Ay' = (y'Kx_1, \ldots, y'Kx_n)$$

Since A is clearly bounded and linear, we know by Remark (i) of III.1.4 that A is compact. Hence there exist $y_1', \ldots, y_m'$ in the 1-ball $S_{Y'}$ of Y' such that for each $y' \in S_{Y'}$, there is a y_j' for which

$$\|Ay' - Ay_j'\| < \frac{\varepsilon}{3}$$

In particular,

$$|y'Kx_i - y_j'Kx_i| < \frac{\varepsilon}{3} \qquad 1 \leq i \leq n \tag{2}$$

From (1) and (2),

$$|K'y'x - K'y_j'x| \leq |y'Kx - y'Kx_i| + |y'Kx_i - y_j'Kx_i| + |y_j'Kx_i - y_j'Kx|$$
$$\leq \|Kx - Kx_i\| + \frac{\varepsilon}{3} + \|Kx_i - Kx\| < \varepsilon$$

Thus $\|K'y' - K'y_j'\| \leq \varepsilon$ whence K' is precompact.

Conversely, suppose K' is compact. Then by what has just been shown, K'' is compact. Now, $K''J_X = J_Y K$, where J_X and J_Y are the natural maps from X into X'' and Y into Y'', respectively. Since J_X is bounded, it is clear that $K''J_X$ is compact and, in particular, $J_Y K$ is precompact. Since J_Y has a bounded inverse, it is easy to see that K is precompact.

If the conjugate of a bounded linear operator is compact, it does not necessarily follow that the operator is compact. Indeed, in Example III.1.7, $T_n \to T$ in $[c_0, Y]$. Thus $T_n' \to T'$ in $[Y', c_0']$. Since T_n is compact, so is T_n'. Hence T' is precompact and therefore compact, since Y' is complete. However, T is not compact.

III.1.12 Corollary. *If the range of a precompact operator is complete, then the range is finite-dimensional.*

Proof. Suppose $B \in [X, Y]$ is precompact and its range Y_1 is complete. Then the operator B_1, considered as the map B from X onto Y_1, is precompact. Furthermore, B_1' has a bounded inverse by Theorem II.3.13. Since B_1' is compact, Theorem III.1.3 asserts that Y_1' is finite-dimensional. Thus Y_1 is finite-dimensional.

III.1.13 Remark. Suppose B is a precompact operator in $[X, Y]$. Since a totally bounded set in a metric space is separable, BS and therefore $\mathcal{R}(B) = \bigcup_{n=1}^{\infty} nBS$ is separable, where S is the 1-ball in X. Thus, in light of Corollary III.1.12, $\mathcal{R}(B)$ is no larger than a proper subspace dense in Y, provided Y is an infinite-dimensional separable Banach space and X is infinite-dimensional. Goldberg and Kruse [3] show that for such X and Y, there always exists a compact operator in $[X, Y]$ whose range is dense in Y. We have also seen in Theorem III.1.3 that B cannot have a bounded inverse on X whenever X is infinite-dimensional. However, in the paper just cited, it is shown that there exists a 1-1 compact operator in $[X, Y]$ if and only if X' contains a countable total subset.

III.2 RELATIONSHIP BETWEEN STRICTLY SINGULAR AND COMPACT OPERATORS

There is a close connection between the class of strictly singular operators and the class of precompact operators. For example, Kato [1], page 288, has shown that these two classes coincide in the space of bounded operators mapping a Hilbert space into a Hilbert space.

Feldman, Gokhberg, and Marcus [1] prove that the only nontrivial closed two-sided ideal in [X], where $X = l_p$, $1 \leq p < \infty$, or $X = c_0$, is the space $\mathcal{K}[X]$ of compact operators. Since it will be shown that the strictly singular operators $\mathcal{S}[X]$ constitute a closed two-sided ideal in $[X]$, it follows that $\mathcal{S}[X] = \mathcal{K}[X]$.

The following two theorems give insight into the connection between the two classes of operators.

III.2.1 Theorem. *Suppose $B \in [X, Y]$. The following three statements are equivalent.*

i. *B is strictly singular.*

ii. *For every infinite-dimensional subspace $M \subset X$, there exists an infinite-dimensional subspace $N \subset M$ such that B is precompact on N.*

iii. *Given $\varepsilon > 0$ and given M an infinite-dimensional subspace of X, there exists an infinite-dimensional subspace $N \subset M$ such that B restricted to N has norm not exceeding ε.*

Proof. (*i*) implies (*ii*). Suppose B is strictly singular and M is an infinite-dimensional subspace of X. Then B_M, the restriction of B to M, is strictly singular. Hence the existence of an N, with the properties asserted in (*ii*), is a consequence of Theorem III.1.9 applied to B_M.

(*ii*) *implies* (*iii*). If M is an infinite-dimensional subspace of X, then B does not have a bounded inverse on any subspace having finite deficiency in M; otherwise, B would be precompact and have a bounded inverse on an infinite-dimensional subspace. But this contradicts Theorem III.1.3. Hence (*iii*) follows by applying Theorem III.1.9 to B_M.

(*i*) follows easily from (*iii*).

The next theorem, due to Lacey [1], implies that if one adds to the above theorem that N has finite deficiency in M, then a characterization of the precompact operators is obtained.

III.2.2 Lemma. *If S is a bounded set in X and $x_1', x_2', \ldots, x_n'$ are in X', then for any $\varepsilon > 0$, there exist $x_1, x_2, \ldots, x_m$ in S such that given $x \in S$, one can find an x_k such that*

$$|x_i'x - x_i'x_k| < \varepsilon \qquad 1 \leq i \leq n$$

Proof. The map A from X to unitary n-space defined by

$$Ax = (x_1'x, x_2'x, \ldots, x_n'x)$$

is bounded and linear. Hence A is compact. Thus there exist $x_1, x_2, \ldots, x_m$ in S such that given $x \in S$, there is an x_k such that

$$\varepsilon > \|Ax - Ax_k\| \geq |x_i'x - x_i'x_k| \qquad 1 \leq i \leq n$$

III.2.3 Theorem. *Suppose $B \in [X, Y]$. Then B is precompact if and only if for every $\varepsilon > 0$ there exists a subspace N having finite deficiency in X such that B restricted to N has norm not exceeding ε.*

Proof. Let B be precompact and let $\varepsilon > 0$ be given. There exist $x_1, x_2, \ldots, x_n$ in the 1-ball S of X such that given $x \in S$, there is an x_k such that

$$\|Bx - Bx_k\| < \frac{\varepsilon}{2} \tag{1}$$

Let $y_1', \ldots, y_n'$ be in the 1-sphere of Y' such that $y_i'Bx_i = \|Bx_i\|$. Now $N = \bigcap_{i=1}^{n} \mathfrak{N}(y_i'B)$ has finite deficiency in X. Suppose $x \in N \cap S$ and x_k is such that (1) holds. Since x is in $\mathfrak{N}(y_k'B)$,

$$\|Bx_k\| = y_k'Bx_k = y_k'(Bx_k - Bx) \leq \|Bx_k - Bx\| < \frac{\varepsilon}{2} \tag{2}$$

Thus, from (1) and (2),

$$\|Bx\| \leq \|Bx - Bx_k\| + \|Bx_k\| < \frac{\varepsilon}{2} + \frac{\varepsilon}{2} \tag{3}$$

Since x was arbitrary in $N \cap S$, (3) implies that B restricted to N has norm not exceeding ε.

Conversely, let $\varepsilon > 0$ be given. By hypothesis, there exists an N of finite deficiency in X such that B_N, the restriction of B to N, has norm not exceeding ε. Since $\|B_N\| = \|B_{\bar{N}}\|$, we may assume that N is closed. Hence, there exist $v_1, v_2, \ldots, v_n$ in X such that

$$X = N \oplus \text{sp}\,\{v_1, \ldots, v_n\}$$

From the construction of the projection in Theorem II.1.16, it follows that there exist $x_1', x_2', \ldots, x_n'$ in X' such that every $x \in X$ has a unique representation of the form

$$x = u + \sum_{i=1}^{n} x_i'(x)v_i \qquad u \in N \tag{4}$$

Since $\|B_N\| \leq \varepsilon$, we have

$$\|Bx\| \leq \varepsilon\|u\| + \sum_{i=1}^{n} |x_i'(x)| \, \|Bv_i\| \tag{5}$$

By (4),

$$\|u\| \leq \|x\| + \sum_{i=1}^{n} |x_i'(x)| \, \|v_i\| \tag{6}$$

Combining (5) and (6) and choosing $K = \max \{\|Bv_i\|, \|v_i\|; 1 \leq i \leq n\}$, it follows that

$$\|Bx\| \leq \varepsilon\|x\| + \varepsilon K \sum_{i=1}^{n} |x_i'(x)| + K \sum_{i=1}^{n} |x_i'(x)| \qquad x \in X \tag{7}$$

Given $\eta > 0$, there exist, by Lemma III.2.2, elements $x_1, x_2, \ldots, x_m$ in S such that given $v \in S$, there is an x_k such that

$$\sum_{i=1}^{n} |x_i'v - x_i'x_k| < \eta$$

Thus, substituting $v - x_k$ for x in (7) and noting that $\|v - x_k\| \leq 2$, we obtain

$$\|Bv - Bx_k\| \leq 2\varepsilon + \varepsilon K\eta + K\eta$$

Since $\varepsilon > 0$ and $\eta > 0$ are arbitrary, it follows that BS is totally bounded.

III.2.4 Theorem. *The set $\mathcal{S}[X, Y]$ of strictly singular operators and the set of precompact operators $\mathcal{PK}[X, Y]$ in $[X, Y]$ are closed subspaces.*

The set of compact operators $\mathcal{K}[X, Y]$ is a subspace (not necessarily closed).

Proof. It is obvious that the sets are closed under scalar multiplication. Suppose K and L are in $\mathcal{K}[X, Y]$. Let $\{x_n\}$ be a bounded sequence in X. Since K is compact, there exists a subsequence $\{y_n\}$ of $\{x_n\}$ such that $\{Ky_n\}$ converges. Since L is compact, there exists a subsequence $\{z_n\}$ of $\{y_n\}$ such that $\{Lz_n\}$ converges. Hence $\{(K + L)z_n\}$ converges, which shows that $K + L$ is compact.

Suppose K and L are in $\mathcal{PK}[X, Y]$. Then K and L are in $\mathcal{K}[X, \hat{Y}]$, where $\hat{Y}$ is the completion of Y. Hence, by the above result, $K + L$ is in $\mathcal{K}[X, \hat{Y}]$, or, equivalently, $K + L$ is in $\mathcal{PK}[X, Y]$.

Suppose K and L are in $\mathcal{S}[X, Y]$. Let M be an infinite-dimensional subspace of X. Then, by Theorem III.2.1, there exists an infinite-dimensional subspace $N_1 \subset M$ such that K is precompact on N_1 and there exists an infinite-dimensional subspace $N \subset N_1$ such that L is precompact on N. Since K is also precompact on N, $K + L$ is precompact on N by what we have just shown. Hence $K + L$ is strictly singular by Theorem III.2.1.

$\mathcal{PK}[X, Y]$ was shown to be closed in Lemma III.1.5. Let $K_n \to K$ in $[X, Y]$, where each K_n is strictly singular. Suppose K has a bounded inverse on subspace $M \subset X$. Then there exists a $c > 0$ such that for all $m \in M$,

$$\|Km\| \geq c\|m\| \tag{1}$$

Choose p so that

$$\|K - K_p\| \leq \frac{c}{2} \tag{2}$$

Then from (1) and (2),

$$\|K_p m\| \geq \|Km\| - \|(K - K_p)m\| \geq \frac{c}{2}\|m\|$$

Thus K_p has a bounded inverse on M. Since K_p is strictly singular, M is finite-dimensional, whence K is strictly singular.

If Y is complete, then the space of precompact operators coincides with the space of compact operators and therefore $\mathcal{K}[X, Y]$ is closed. Example III.1.7 showed that if Y is not required to be complete, then $\mathcal{K}[X, Y]$ need not be closed.

III.2.5 Theorem. *Let Z be a normed linear space. If K is in $\mathcal{K}[X, Y]$, $\mathcal{PK}[X, Y]$, or $\mathcal{S}[X, Y]$, then for $A \in [Z, X]$, KA is in $\mathcal{K}[Z, Y]$, $\mathcal{PK}[Z, Y]$, or $\mathcal{S}[Z, Y]$, respectively. Similarly, if B is in $[Y, Z]$, then BK is in $\mathcal{K}[X, Z]$, $\mathcal{PK}[X, Z]$, or $\mathcal{S}[X, Z]$, respectively.*

Proof. We shall only prove the theorem for $K \in \mathcal{S}[X, Y]$. The proofs for K in $\mathcal{K}[X, Y]$ or $\mathcal{PK}[X, Y]$ are also easy. Suppose KA has a bounded inverse on a subspace M of Z. Then there exists a $c > 0$ such that for all $x \in M$

$$\|KAx\| \geq c\|x\| \geq \frac{c}{\|A\|}\|Ax\|$$

Thus K has a bounded inverse on AM, and therefore AM is finite-dimensional. But A is 1-1 since KA is 1-1. Hence M is finite-dimensional. Consequently, KA is strictly singular.

Suppose BK has a bounded inverse on a subspace N of X. Then there exists a $c > 0$ such that for all $x \in N$,

$$\|B\| \|Kx\| \geq \|BKx\| \geq c\|x\|$$

Thus K has a bounded inverse on N, whence N is finite-dimensional. Therefore BK is strictly singular.

III.3 EXAMPLES OF NONCOMPACT, STRICTLY SINGULAR OPERATORS

For the purpose of exhibiting some strictly singular operators, we prove Theorem III.3.4 and Corollary III.3.6, which are due to Whitley [1].

III.3.1 Definition. *A linear operator which takes bounded sequences onto sequences which have a weakly convergent subsequence is called* ***weakly compact.***

III.3.2 Remark. Every weakly compact operator is bounded. For suppose B is a weakly compact operator which is unbounded. Then there exists a sequence $\{x_n\}$ such that $\{Bx_n\}$ converges weakly and $\|Bx_n\| \to \infty$. But this contradicts Theorem II.1.12.

III.3.3 Theorem. *If at least X or Y is reflexive, then every operator in $[X, Y]$ is weakly compact.*

Proof. The proof follows easily from Theorem I.6.15 and the continuity of the operator.

III.3.4 Theorem. *A weakly compact operator on a Banach space which maps weakly convergent sequences onto norm convergent sequencies is strictly singular.*

Proof. Let $B \in [X, Y]$ satisfy the hypotheses of the theorem. Suppose B has a bounded inverse on a closed subspace $M \subset X$. Let $\{m_k\}$ be a sequence in the 1-ball of M. Then there exists a subsequence $\{v_k\}$ of $\{m_k\}$ such that $\{Bv_k\}$ converges weakly to some $y \in Y$. Since BM is closed, y is in BM; otherwise there exists a $y' \in Y'$ such that $y'y = 1$ and $y'(BM) = 0$, implying that $\{Bv_k\}$ does not converge weakly to y. From the assumption that B has a bounded inverse on M, it follows that $\{v_k\}$ converges weakly to $B_M{}^{-1}y$. From the hypothesis, $\{Bv_k\}$ converges in norm to y and therefore $\{v_k\}$ converges in norm to $B_M{}^{-1}y$. Hence the 1-ball of M is compact and M is finite-dimensional. Therefore B is strictly singular.

The proof of the next theorem requires quite a bit of integration theory and will not be included here. The reader is referred to Dunford and Schwartz [1], Theorem VI.8.12, the remark preceding the theorem on page 508, the remark following Theorem VI.8.14 on pages 510–511, and the representation of the conjugate spaces in tables on pages 374–379. We cite only a few spaces X for which the following theorem holds.

III.3.5 Theorem. *Every weakly compact map from X or X' into a Banach space takes weakly convergent sequences onto norm convergent sequences whenever X is any one of the following Banach spaces.*

i. $\mathfrak{L}_1(S, \Sigma, \mu)$ *and* $\mathfrak{L}_\infty(S, \Sigma, \mu)$, *where* (S, Σ, μ) *is a positive measure space.*
ii. C(S), *where S is a compact Hausdorff space.*
iii. B(S), *where S is an arbitrary set.*

Thus every weakly compact operator from X or X' into a Banach space is strictly singular.

Pelczynski [2] proved that the *strictly singular operators* and the *weakly compact operators* are the same in the following spaces:

i. $[C(S), Y]$, Y an arbitrary Banach space and S a compact Hausdorff space.
ii. $[\mathfrak{L}_1(S, \Sigma, \mu_1), \mathfrak{L}_1(S, \Sigma, \mu_2)]$, S a topological space, Σ the field of all Borel subsets of S, and μ_1 and μ_2 nontrivial measures (for details of these spaces, see Dunford and Schwartz [1], Chap. IV).

III.3.6 Corollary. *Let X be any one of the spaces in the above theorem. Then every bounded linear map from X or X' into a reflexive space is strictly singular.*

Proof. Theorems III.3.3 and III.3.5.

The rest of this section gives examples of strictly singular operators which are not compact.

III.3.7 Example. Let $X = l_1$ and let Y be any infinite-dimensional separable reflexive Banach space. By Corollary II.4.5, there exists a bounded linear operator B which maps X onto Y. Hence B is strictly singular but not compact by Corollaries III.3.6 and III.1.12. Without referring to Corollary III.3.6, B can be shown to be strictly singular based on the fact that weak convergence is the same as norm convergence in l_1, as remarked in I.6.14. B' is not strictly singular, since it has a bounded inverse by Theorem II.4.4. Now B'' maps X'' into reflexive space Y'', and X'' may be identified with the conjugate of l_∞. Thus, by Corollary III.3.6, B'' is strictly singular. This example shows, in particular, that the conjugate of a strictly singular operator need not be strictly singular and, conversely, the conjugate of a bounded nonstrictly singular operator can be strictly singular.

The following theorem, which we do not prove, is needed for the next two examples of integral operators which are strictly singular but not compact. The proof of the theorem appears in Dunford and Schwartz [1], Corollary IV.8.11, and in Zaanen [1], pages 324–325.

III.3.8 Theorem. *Let (S, Σ, μ) be a positive measure space with $\mu(S) < \infty$. A sequence $\{y_n\}$ in $\mathfrak{L}_1(S, \Sigma, \mu)$ has a weakly convergent subsequence if and only if it is bounded in $\mathfrak{L}_1(S, \Sigma, \mu)$ and $\int_E y_n \, d\mu$ converges to zero uniformly for all y_n as $\mu(E) \to 0$.*

III.3.9 Corollary. *Let k be bounded and measurable on $[0, 1] \times [0, 1]$. Define K as a map from $\mathfrak{L}_1([0, 1])$ into $\mathfrak{L}_p([0, 1])$, $1 \leq p < \infty$, by*

$$(Kf)(t) = \int_0^1 k(s, t) f(s) \, ds$$

Then K is weakly compact.

Proof. For $f \in \mathfrak{L}_1([0, 1])$ and E a measurable subset of $[0, 1]$,

$$\text{(1)} \qquad \left[\int_E |(Kf)(t)|^p \, dt \right]^{1/p} \leq (\mu(E))^{1/p} \sup_{0 \leq s,t \leq 1} |k(s, t)| \int_0^1 |f(s)| \, ds$$

Taking $E = [0, 1]$ and $1 < p < \infty$, (1) shows that K is bounded and therefore weakly compact by Theorem III.3.3. For $p = 1$, let $\{f_n\}$ be a bounded sequence in $\mathfrak{L}_1([0, 1])$. Then, by (1), $\{Kf_n\}$ is bounded in $\mathfrak{L}_1([0, 1])$ and $\int_E |(Kf_n)(t)| \, dt$ converges to zero uniformly for all f_n as $\mu(E) \to 0$. It follows from Theorem III.3.8 that K is weakly compact.

The next two examples exhibit integral operators which are strictly singular but not compact.

III.3.10 Example. For $n = 1, 2, \ldots$, let k be defined on $[0, 1] \times [0, 1]$ by

$$k(s, t) = \begin{cases} 0 & (s, t) \in \left(\frac{1}{2^n}, \frac{1}{2^{n-1}} \right] \times \left(\frac{2j}{2^n}, \frac{2j+1}{2^n} \right], \\ & \qquad j = 0, 1, \ldots, 2^{n-1} - 1 \\ 2 & (s, t) \in \left(\frac{1}{2^n}, \frac{1}{2^{n-1}} \right] \times \left(\frac{2j+1}{2^n}, \frac{2j+2}{2^n} \right], \\ & \qquad j = 0, 1, \ldots, 2^{n-1} - 1 \\ 0 & \text{when } s = 0 \text{ or } t = 0 \end{cases}$$

Clearly k is bounded and measurable. Define the corresponding integral operator $K:\mathfrak{L}_1([0, 1]) \to \mathfrak{L}_p([0, 1]), 1 \leq p < \infty$, by

$$(Kf)(t) = \int_0^1 k(s, t)f(s)\, ds$$

Then K is strictly singular by Corollary III.3.9 and Theorem III.3.5. However, K is not compact. To see this, define sequence $\{f_n\}$ by

$$f_n(s) = \begin{cases} 2^n & s \in \left(\frac{1}{2^n}, \frac{1}{2^{n-1}}\right], \ n = 1, 2, \ldots \\ 0 & \text{otherwise} \end{cases}$$

Then $\|f_n\|_1 = 1$ and a simple calculation shows that

$$\|Kf_n - Kf_m\|_p \geq \|Kf_n - Kf_m\|_1 = 1 \qquad n \neq m$$

where $\| \quad \|_p$ is the norm on $\mathfrak{L}_p(I)$. Hence $\{Kf_n\}$ has no convergent subsequence in $\mathfrak{L}_p(I)$.

III.3.11 Remarks. It is shown in Dunford and Schwartz [1], Theorem VI.4.8, that the conjugate of a weakly compact operator is weakly compact. Hence, taking the operator K in the above example as a map from $\mathfrak{L}_1([0, 1])$ to $\mathfrak{L}_1([0, 1])$, K' is also strictly singular by Theorem III.3.5.

III.3.12 Example. Let k be bounded and measurable on $[0, 1] \times [0, 1]$ and let φ be in $\mathfrak{L}_1([-1, 1])$. It follows from Theorem 0.11 that the operator K defined by

$$(Kf)(t) = \int_0^1 k(s, t)\varphi(t - s)f(s)\, ds$$

is bounded as a map from $\mathfrak{L}_1(I)$ into $\mathfrak{L}_1(I)$, $I = [0, 1]$. To show that K is strictly singular, it suffices, by Theorem III.3.5, to show that K is weakly compact. Let E be a measurable subset of I. Then for $f \in \mathfrak{L}_1(I)$ and $M = \sup_{0 \leq s,t \leq 1} |k(s, t)|$, Fubini's theorem gives

$$(1) \quad \int_E |(Kf)(t)|\, dt \leq M \int_E dt \int_I |\varphi(t - s)|\, |f(s)|\, ds$$

$$= M \int_I \left(\int_{E-s} |\varphi(t)|\, dt \right) |f(s)|\, ds$$

Since φ is integrable on $[-1, 1]$, there exists a $\delta > 0$, depending on $\varepsilon > 0$, such that whenever E has meaure less than δ,

$$\int_{E-s} |\varphi(t)|\, dt < \varepsilon \qquad s \in I$$

For such E, (1) implies that

$$\int_E |(Kf)(t)|\,dt \leq M\varepsilon \qquad \|f\| \leq 1$$

Hence K is weakly compact by Theorem III.3.8.

As a particular case of this example we see that for g bounded and measurable on $[0, 1] \times [0, 1]$, the operator K_1 defined by

$$(K_1 f)(t) = \int_0^1 \frac{g(s, t)}{|t - s|^\alpha} f(s)\,ds \qquad 0 < \alpha < 1$$

is strictly singular as a map from $\mathfrak{L}_1([0, 1])$ to $\mathfrak{L}_1([0, 1])$. Upon choosing

$$g(s, t) = k(s, t)|t - s|^\alpha \qquad 0 < \alpha < 1$$

where k is given in Example III.3.10, $K_1 = K$, which is not compact.

Feldman, Gokhberg, and Marcus [1] show that the compact operators and the strictly singular operators on $\mathfrak{L}_p([0, 1])$, $1 \leq p < \infty$, $p \neq 2$, are not the same. For additional examples of noncompact, strictly singular operators, the reader is referred to Goldberg and Thorp [1] and Lacey and Whitley [1].

III.4 CONJUGATES OF STRICTLY SINGULAR OPERATORS

In this section we consider when the conjugate of a strictly singular operator is strictly singular and, conversely, when the strict singularity of the conjugate implies the strict singularity of the operator. The results given here are due to Whitley [1].

III.4.1 Definition. *A normed linear space Y is called* ***subprojective*** *if for every closed infinite-dimensional subspace W of Y, there exists a closed infinite-dimensional subspace W_1 of W and a projection from Y onto W_1.*

A Hilbert space is subprojective.

The proof of the next theorem may be found in Pelczynski [1].

III.4.2 Theorem. *The spaces l_p, $1 \leq p < \infty$, $\mathfrak{L}_p([0, 1])$, $2 \leq p < \infty$, and c_0 are subprojective.*

III.4.3 Theorem. *Let X and Y be complete and let Y be subprojective. If the conjugate of $T \in [X, Y]$ is strictly singular, then T is strictly singular.*

Proof. Assume T is not strictly singular. Then there exists an infinite-dimensional closed subspace M of X such that the restriction T_M of T to M has a bounded inverse. Since TM is a closed infinite-dimensional subspace of Y, the subprojectivity of Y implies the existence of an infinite-dimensional closed subspace V of TM and a projection P from Y onto $V = TN$, where $N = T_M^{-1}V \subset M$. Since P is surjective, P' has a bounded inverse on the conjugate space $(TN)'$ by Theorem II.3.13. We show that the strictly singular operator T' has a bounded inverse on the infinite-dimensional subspace $P'((TN)')$, which is a contradiction resulting from the assumption that T is not strictly singular. Suppose $v' \epsilon (TN)'$ and x is in the 1-sphere of N. Then

$$\|T'(P'v')\| \geq |T'(P'v')x| = |v'PTx| = |v'Tx| \tag{1}$$

Since T_M has a bounded inverse and N is contained in M, it follows that for $\|x\| = 1$ and $x \epsilon N$,

$$|v'Tx| \geq \frac{|v'Tx|}{\|T_M^{-1}\| \, \|Tx\|} \tag{2}$$

As a consequence of (1) and (2), we get

$$\|T'(P'v')\| \geq \frac{\|v'\|}{\|T_M^{-1}\|} \geq \frac{\|P'v'\|}{\|P'\| \, \|T_M^{-1}\|} \qquad v' \epsilon (TN)'$$

Thus T' has a bounded inverse on $P'((TN)')$.

III.4.4 Corollary. *Suppose that X is reflexive and that X' is subprojective. If $T \epsilon [X, Y]$ is strictly singular, then T' is also strictly singular.*

Proof. Since $T'' = J_Y T J_X^{-1}$, where J_X and J_Y are the natural maps from X into X'' and Y into Y'', respectively, T'' is strictly singular by Theorem III.2.5. Since T' is in $[Y', X']$ and X' is subprojective, T' is strictly singular.

III.4.5 Corollary. *Let T be in $[X, Y]$. If X is a Hilbert space and T is strictly singular, then so is T'. If Y is a Hilbert space and T' is strictly singular, then so is T.*

Proof. By Theorem I.7.21, the conjugate of a Hilbert space is a Hilbert space. Since a Hilbert space is subprojective, the corollary follows from Corollary III.4.4 and Theorem III.4.3.

chapter IV

OPERATORS WITH CLOSED RANGE

Theorem IV.1.2 provides the motivation for the study of closed operators with closed range. There are many important applications of the theorem. For example, given a differential operator T defined on some subspace of $\mathfrak{L}_p(\Omega)$, one is interested in determining the family of functions $y \in \mathfrak{L}_q(\Omega)$ for which $Tf = y$ has a solution. If it is known that T is closed and has a closed range (examples of such differential operators are given in Chaps. VI and VII), then the space of such y is the orthogonal complement of the solutions to the homogeneous equation $T'g = 0$. This information is very useful since oftentimes $\mathfrak{N}(T')$ is finite-dimensional.

We shall also see the vital role that closed operators with closed range play in perturbation theory.

As before, T is a linear operator with domain in normed linear space X (not necessarily dense in X) and range in normed linear space Y. The spaces are assumed complete and the operators are assumed closed only when specifically stated.

IV.1 MINIMUM MODULUS OF AN OPERATOR

IV.1.1 Lemma. *Let X and Y be complete and let T be closed. Then T has a bounded inverse if and only if T is 1-1 and has a closed range.*

Proof. Suppose T has a bounded inverse. Replacing T' by T in the proof of Theorem II.3.1 shows that $\mathcal{R}(T)$ is closed. Conversely, suppose T is 1-1 and $\mathcal{R}(T)$ is closed. Then T^{-1} is a closed operator from Banach space $\mathcal{R}(T)$ into Banach space X. Hence T^{-1} is continuous by the closed-graph theorem.

IV.1.2 Theorem. *Let X and Y be complete. Suppose T is closed with domain dense in X. Then the following statements are equivalent.*

i. $\mathcal{R}(T)$ *is closed.*
ii. $\mathcal{R}(T')$*is closed.*
iii. $\mathcal{R}(T')$ *is the orthogonal complement of* $\mathfrak{N}(T)$*; that is,*

$$\mathcal{R}(T') = \mathfrak{N}(T)^{\perp}$$

iv. $\mathcal{R}(T)$ *is the orthogonal complement of* $\mathfrak{N}(T')$*; that is,*

$$\mathcal{R}(T) = {}^{\perp}\mathfrak{N}(T')$$

Proof. Theorem II.3.7 implies that (*i*) is equivalent to (*iv*).

(*i*) implies (*iii*). Let $\hat{T}$ be the 1-1 operator induced by T. Then $\hat{T}$ has a bounded inverse by Lemma IV.1.1 and therefore $\mathcal{R}((\hat{T})') = (X/\mathfrak{N}(T))'$ by Theorem II.4.4. Now, by (*ii*) of Theorem I.6.4, $(X/\mathfrak{N}(T))'$ is equivalent to $\mathfrak{N}(T)^{\perp}$ under the map V given by $(Vz')x = z'[x]$. Thus

$$V(\mathcal{R}(\hat{T})') = \mathfrak{N}(T)^{\perp} \tag{1}$$

Since $\mathfrak{D}(T') = \mathfrak{D}((\hat{T})')$ by Lemma II.4.7 and since $(\hat{T})'y'[x] = T'y'x$, $y' \in \mathfrak{D}(T')$, it follows from (1) and the definition of V that

$$\mathcal{R}(T') = V(\mathcal{R}(\hat{T})') = \mathfrak{N}(T)^{\perp}$$

(*ii*) *is an immediate consequence of* (*iii*).

(*ii*) *implies* (*i*). Let $Y_1 = \overline{\mathcal{R}(T)}$ and let T_1 be the operator T considered as a map into Banach space Y_1. Obviously, T_1 is closed. To prove (*i*), it need only be shown that $\mathcal{R}(T_1) = Y_1$. By Theorem II.4.4, it sufficies to prove that T_1' has a bounded inverse. Since $\overline{\mathcal{R}(T_1)} = Y_1$, T_1' has an inverse by Theorem II.3.7. Furthermore, $\mathcal{R}(T_1')$ is easily seen to be $\mathcal{R}(T')$. Hence, T_1' has a bounded inverse by Lemma IV.1.1.

To each linear operator T having a closed kernel we associate a number $\gamma(T)$, which plays a vital role in subsequent sections. The number arises from the following considerations.

Suppose T is closed and X and Y are complete. Let $\hat{T}$ be the 1-1 operator induced by T. Then, by Lemma IV.1.1, $\mathcal{R}(T) = \mathcal{R}(\hat{T})$ is closed if and only if $\hat{T}$ has a bounded inverse or, equivalently,

$$0 < \inf \left\{ \frac{\|\hat{T}[x]\|}{\|[x]\|} \,\middle|\, [x] \in \mathfrak{D}(\hat{T}),\, x \notin \mathfrak{N}(T) \right\}$$

$$= \inf \left\{ \frac{\|Tx\|}{d(x, \mathfrak{N}(T))} \,\middle|\, x \in \mathfrak{D}(T),\, x \notin \mathfrak{N}(T) \right\} = \frac{1}{\|(\hat{T})^{-1}\|}$$

IV.1.3 Definition. *Let $\mathfrak{N}(T)$ be closed. The minimum modulus of T, written $\gamma(T)$, is defined by*

$$\gamma(T) = \inf_{x \in \mathfrak{D}(T)} \frac{\|Tx\|}{d(x, \mathfrak{N}(T))}$$

where $0/0$ is defined to be ∞.

IV.1.4 Example. Let $X = \mathfrak{L}_p(I)$ and $Y = \mathfrak{L}_q(I)$, where $I = [a, b]$ is compact and $1 \leq p, q \leq \infty$. Define T as follows.

$$\mathfrak{D}(T) = \{f \mid f^{(n-1)} \text{ exists and is absolutely continuous on } I,\ f^{(n)} \in Y\}$$
$$Tf = f^{(n)}$$

(Recall that an absolutely continuous function is differentiable almost everywhere.)

$\mathfrak{N}(T)$ is the space of polynomials of degree at most $n - 1$. Successive integration by parts shows that given $f \in \mathfrak{D}(T)$, there exists a $z \in \mathfrak{N}(T)$ such that

$$f(t) = z(t) + \int_a^t \frac{(t-s)^{n-1}}{(n-1)!} f^{(n)}(s)\, ds \tag{1}$$

Thus, for $1 < q \leq \infty$, Hölder's inequality gives

$$|f(t) - z(t)| \leq \frac{\|f^{(n)}\|_q}{(n-1)!} \left(\int_a^t (t-s)^{(n-1)q'}\, ds \right)^{1/q'} \tag{2}$$

$$= \frac{\|f^{(n)}\|_q (t-a)^{n-1+1/q'}}{(n-1)![(n-1)q' + 1]^{1/q'}}$$

where $\| \ \|_q$ is the norm on $\mathfrak{L}_q(I)$ and q' is conjugate to q. For $q = 1$, (1) implies that

$$|f(t) - z(t)| \leq \frac{\|f^{(n)}\|_1 (t-a)^{n-1}}{(n-1)!} \tag{3}$$

If we agree to write $1/\infty = 0$, $\infty^0 = 1$, and $\infty/\infty = 1$, then it follows from (2) and (3) that for $1 \leq p, q \leq \infty$,

$$(4) \quad \|f - z\|_p \leq \frac{\|f^{(n)}\|_q (b-a)^{n-1+1/q'+1/p}}{(n-1)![(n-1)q'+1]^{1/q'}[(n-1)p+p/q'+1]^{1/p}}$$

Since z is in $\mathfrak{N}(T)$, we may conclude from (4) that

$$\gamma(T) \geq \frac{(n-1)![(n-1)q'+1]^{1/q'}[(n-1)p+1+p/q']^{1/p}}{(b-a)^{n-1+1/q'+1/p}}$$

where $1 \leq p, q \leq \infty$, $1/\infty \equiv 0$, $\infty^0 \equiv 1$, and $\infty/\infty \equiv 1$.

In particular, if $p = q = 2$, then

$$\gamma(T) \geq \frac{(n-1)!(2n-1)^{\frac{1}{2}}(2n)^{\frac{1}{2}}}{(b-a)^n}$$

An estimate for $\gamma(T)$ is given in Chap. VI for a rather large class of ordinary differential operators T.

IV.1.5 Example. Take $X = Y = l_p$, $1 \leq p \leq \infty$. Let $\{\lambda_k\}$ be a bounded sequence of numbers and let T be defined on all of X by

$$T(\{\alpha_k\}) = \{\lambda_k \alpha_k\}$$

For each scalar λ we compute $\gamma(T_\lambda)$, where $T_\lambda = \lambda I - T$ and I is the identity operator.

Define E_λ to be the set of integers k for which $\lambda \neq \lambda_k$. Then $\mathfrak{N}(T_\lambda)$ is the set of elements of the form $(\beta_1, \beta_2, \ldots) \in l_p$, where β_j is arbitrary if $j \notin E_\lambda$ and $\beta_k = 0$ if $k \in E_\lambda$. For simplicity we compute $\gamma(T_\lambda)$ when $X = l_1$. The computation for $1 < p \leq \infty$ is essentially the same. Let $x = \{\eta_k\}$ be in X. Then for $y = \{\beta_k\} \in \mathfrak{N}(T_\lambda)$,

$$\|x - y\| = \sum_{k \notin E_\lambda} |\eta_k - \beta_k| + \sum_{k \in E_\lambda} |\eta_k|$$

Since β_k can be chosen to be η_k for $k \notin E_\lambda$, it follows that

$$\|[x]\| = d(x, \mathfrak{N}(T_\lambda)) = \sum_{k \in E_\lambda} |\eta_k|$$

Therefore,

$$(1) \qquad \frac{\|T_\lambda x\|}{\|[x]\|} = \frac{\sum_{k \in E_\lambda} |\lambda - \lambda_k|\,|\eta_k|}{\sum_{k \in E_\lambda} |\eta_k|} \geq \inf_{k \in E_\lambda} |\lambda - \lambda_k|$$

which implies that

$$\text{(2)} \qquad \gamma(T_\lambda) \geq \inf_{k \in E_\lambda} |\lambda - \lambda_k| = m$$

On the other hand, given $\varepsilon > 0$, there exists some $i \in E_\lambda$ such that $|\lambda - \lambda_i| < m + \varepsilon$. Let u_i be the element in l_1 all of whose terms are zero except the ith term which is 1. Then from (1) and (2),

$$m \leq \gamma(T_\lambda) \leq \frac{\|T_\lambda u_i\|}{\|[u_i]\|} = |\lambda - \lambda_i| < m + \varepsilon$$

Since $\varepsilon > 0$ was arbitrary,

$$\inf_{k \in E_\lambda} |\lambda - \lambda_k| = \gamma(\lambda I - T)$$

The next theorem was proved in the discussion preceding Definition IV.1.3.

IV.1.6 Theorem. *Let X and Y be complete and let T be closed. Then T has a closed range if and only if $\gamma(T) > 0$.*

Theorem IV.1.6, together with the next theorem, shows that if X and Y are complete, then any two of the following three statements imply the other.

$$T \text{ closed} \qquad \mathcal{R}(T) \text{ closed} \qquad \gamma(T) > 0$$

IV.1.7 Theorem. *If $\gamma(T) > 0$ and $\mathcal{R}(T)$ is closed, then T is closed.*

Proof. Let $\hat{T}$ be the 1-1 operator induced by T. Then $(\hat{T})^{-1}$ is continuous with domain a closed subspace of Y. Thus $(\hat{T})^{-1}$ is closed, and therefore $\hat{T}$ is also closed. Hence T is closed by Lemma II.4.7.

IV.1.8 Theorem. *Let $\mathfrak{N}(T)$ be closed and let $\mathfrak{D}(T)$ be dense in X. If $\gamma(T) > 0$, then $\gamma(T) = \gamma(T')$ and T' has a closed range.*

Proof. Let $Y_1 = \mathcal{R}(T)$ and let T_1 be the operator T considered as a map onto Y_1. The 1-1 operator $\hat{T}_1$ is surjective and has a bounded inverse, since $\gamma(T_1) = \gamma(T) > 0$. Hence, by the State diagram II.3.14, $(\hat{T}_1)'$ is also in I_1. Before evaluating $\gamma(T')$, we need the following observations.

i. By Theorems II.3.7 and I.6.4, $Y'/\mathfrak{N}(T') = Y'/\mathcal{R}(T)^\perp$ and $Y'/\mathcal{R}(T)^\perp$ is equivalent to Y_1' under the map $[y'] \to y_R'$, where y_R' is the restriction of y' to Y_1. In particular, $\|[y']\| = \|y_R'\|$.

ii. It is clear that $\|(\hat{T})'y'\| = \|(\hat{T}_1)'y'_R\|$. Thus, for $y' \in \mathfrak{D}(T')$, $\|(\hat{T}_1)'y'_R\| = \|T'y'\|$ by Lemma II.4.7.

iii. By Theorems II.3.9 and II.2.8, $\|((\hat{T}_1)')^{-1}\| = \|(\hat{T}_1^{-1})'\| = \|\hat{T}_1^{-1}\|$.

From (*i*), (*ii*), and (*iii*) it follows that

$$\gamma(T') = \inf_{[y'] \in \mathfrak{D}(\widehat{T'})} \frac{\|T'y'\|}{\|[y']\|} = \inf_{y_R' \in Y_1'} \frac{\|(\hat{T}_1)'y'_R\|}{\|y'_R\|} = \frac{1}{\|((\hat{T}_1)')^{-1}\|}$$

$$= \frac{1}{\|\hat{T}_1^{-1}\|} = \gamma(T)$$

Thus T' has a closed range since $\gamma(T') > 0$ and T' is closed.

The above theorem does not hold if $\gamma(T)$ is not required to be positive. Indeed, Diagram II.3.14 shows that state $(\mathrm{I}_2, \mathrm{III}_1)$ can exist. However, if T is assumed to be closed, then the following corollary due to Kato [1] is obtained.

IV.1.9 Corollary. *If X and Y are complete and T is closed, then* $\gamma(T) = \gamma(T')$.

Proof. By Theorems IV.1.6 and IV.1.2, the statements $\gamma(T) = 0$, $\mathfrak{R}(T)$ is not closed, $\mathfrak{R}(T')$ is not closed, and $\gamma(T') = 0$ are all equivalent. If $\gamma(T) > 0$, then by the theorem just proved, $\gamma(T) = \gamma(T')$.

The remaining portion of the section gives some conditions under which a closed operator has a closed range.

IV.1.10 Theorem. *Let X and Y be complete and let T be closed. A sufficient condition that T have a closed range is that T maps bounded closed sets onto closed sets. If $\mathfrak{N}(T)$ is finite-dimensional, then the condition is also necessary.*

Proof. Suppose T maps bounded closed sets onto closed sets. If $\mathfrak{R}(T)$ were not closed, then $\gamma(T) = 0$, which implies the existence of a sequence $\{[x_n]\}$ in $\mathfrak{D}(T)/\mathfrak{N}(T)$ such that

$$(1) \qquad \|[x_n]\| = 1 \qquad \text{and} \qquad Tx_n \to 0$$

We show that the existence of such a sequence leads to a contradiction. Consequently, $\mathfrak{R}(T)$ has to be closed.

Let $\{z_n\}$ be a sequence such that $z_n \in [x_n]$ and $\|z_n\| \leq 2$. Then $Tz_n = Tx_n \to 0$. If $\{z_n\}$ has no convergent subsequence, it is a closed bounded set. Thus, by hypothesis, $\{Tz_n\}$ is a closed set, whence $Tz_N = 0$ for some N. But then $\|[x_N]\| = \|[z_N]\| = 0$, which contradicts (1). If $z_{n'} \to z$ for some subsequence $\{z_{n'}\}$, then $Tz = 0$, since $Tz_{n'} \to 0$ and T is closed. Hence

$$[x_{n'}] = [z_{n'}] \to [z] = [0]$$

which again contradicts (1).

Suppose $\mathfrak{N}(T)$ is finite-dimensional and $\mathfrak{R}(T)$ is closed. Let S be a closed bounded set in X and let $Tx_n \to y$, where $\{x_n\}$ is a sequence in S. Since T has a closed range, $y = Tx$ for some $x \in \mathfrak{D}(T)$ and $\hat{T}$, the 1-1 operator induced by T, has a bounded inverse. Thus

$$[x_n] = \hat{T}^{-1}Tx_n \to \hat{T}^{-1}Tx = [x]$$

Consequently, there exists a sequence $\{z_n\}$ in $\mathfrak{N}(T)$ such that $x_n + z_n \to x$. Since $\{x_n\}$ is a sequence in the bounded set S, $\{z_n\}$ is also bounded. The finite dimensionality of $\mathfrak{N}(T)$ assures the existence of a subsequence $\{z_{n'}\}$ which converges to an element $z \in \mathfrak{N}(T)$. Hence $x_{n'} \to x - z$. Since S is closed and T is a closed operator, it follows that $x - z$ is in $\mathfrak{D}(T) \cap S$ and $y = Tx = T(x - z) \in TS$. Thus TS is closed.

IV.1.11 Lemma. *Let Y be complete. Suppose that M and N are linearly independent closed subspaces of Y such that $Y = M \oplus N$* (cf. Definition I.7.16). *Then $Y' = M^\perp \oplus N^\perp$.*

Proof. Clearly, $M^\perp \cap N^\perp = (0)$. By Theorem II.1.14, there exists a projection P from Y onto M with $\mathfrak{N}(P) = N$. Given $y' \in Y'$, let $y_1' = y'P$ and let $y_2' = y'(I - P)$. Then $y' = y_1' + y_2'$, $y_1' \in N^\perp$ and $y_2' \in M^\perp$.

IV.1.12 Theorem. *Let X and Y be complete and let T be closed. Suppose N is a closed subspace of Y such that $\mathfrak{R}(T) \oplus N$ is closed. Then $\mathfrak{R}(T)$ is closed. If $\mathfrak{D}(T)$ is dense in X, then $T'Y' = T'N^\perp$.*

Proof. Define T_0 by

$$\mathfrak{D}(T_0) = \mathfrak{D}(T) \times N \subset X \times Y \qquad T_0(x, n) = Tx + n$$

The linear operator T_0 is closed. Suppose $(x_k, n_k) \to (x, y) \in X \times Y$ and $T_0(x_k, n_k) \to z$. Then $x_k \to x$, $n_k \to y$ and $Tx_k + n_k \to z$. Hence $y \in N$ and $Tx_k \to z - y$. Since T is closed, x is in $\mathfrak{D}(T)$ and $Tx = z - y$. Thus $(x, y) \in \mathfrak{D}(T) \times N = \mathfrak{D}(T_0)$ and $T_0(x, y) = Tx + y = z$. By hypothesis, $\mathfrak{R}(T_0) = \mathfrak{R}(T) \oplus N$ is closed. Consequently, $\gamma(T_0) > 0$. To prove that $\mathfrak{R}(T)$ is closed, we show that $\gamma(T) > 0$. Since $\mathfrak{R}(T) \cap N = (0)$, it is clear that $\mathfrak{N}(T_0) = \mathfrak{N}(T) \times \{0\}$. Therefore, given $x \in \mathfrak{D}(T)$,

$$\|Tx\| = \|T_0(x, 0)\| \geq \gamma(T_0)\, d((x, 0), \mathfrak{N}(T_0)) = \gamma(T_0)\, d(x, \mathfrak{N}(T))$$

Hence $\gamma(T) \geq \gamma(T_0) > 0$.

Suppose $\mathfrak{D}(T)$ is dense in X and $y' \in \mathfrak{D}(T')$ is given. Let

$$Y_1 = \mathfrak{R}(T) \oplus N$$

Then, by Lemma IV.1.11,

$$Y_1' = \mathfrak{R}(T)^0 \oplus N^0$$

where $\mathfrak{R}(T)^0$ and N^0 are the orthogonal complements in Y_1' of $\mathfrak{R}(T)$ and N, respectively. Thus, if y_R' is y' restricted to Y_1, there exist $y_1' \in \mathfrak{R}(T)^0$ and $y_2' \in N^0$ such that $y_R' = y_1' + y_2'$. Let v' be a continuous linear extension of y_1' to all of Y. Then v' is in $\mathfrak{R}(T)^\perp$, and therefore v' is in $\mathfrak{N}(T')$ by Theorem IV.1.2. Hence $T'y' = T'(y' - v')$. Moreover, $y' - v'$ is in $N^\perp$, since $y' - v'$ is y_2' on N.

IV.1.13 Corollary. *Let X and Y be complete and let T be closed. If $\mathfrak{R}(T)$ has finite deficiency in Y, then $\mathfrak{R}(T)$ is closed.*

Proof. $Y = \mathfrak{R}(T) \oplus N$, where N is some finite-dimensional subspace of Y.

IV.1.14 Remarks. The above corollary need not hold if T is not closed. To show this, we first make the following observations.

i. Every linear operator defined on a finite-dimensional normed linear space V is continuous. This can be shown by first taking V to be unitary n-space U^n and then using the fact that U^n is isomorphic to V.

ii. Every linear functional f on X with closed kernel is continuous. This follows from the fact that the 1-1 operator $\hat{f}$ induced by f maps the one-dimensional space $X/\mathfrak{N}(f)$ (assuming $f \neq 0$) onto the scalars. Hence $\hat{f}$ is continuous and therefore f is continuous by Lemma II.4.7.

iii. There exists an unbounded linear functional on X, provided X is infinite-dimensional. Take $\{v_1, v_2, \ldots\}$ to be an infinite subset of a Hamel basis $\{x_\alpha\}$ of X, where $\|x_\alpha\| = 1$. Let f be the linear functional on X defined by $f(x_\alpha) = 0$ when $x_\alpha \neq v_k$, $1 \leq k$, and $f(v_k) = k$. Then f is clearly unbounded.

By the above results, $\mathfrak{N}(f)$ is not closed and $X = \mathfrak{N}(f) + \operatorname{sp}\{x\}$, $x \notin \mathfrak{N}(f)$. The linear operator T defined on X by $T(u + \alpha x) = u$, $u \in \mathfrak{N}(f)$, has range $\mathfrak{N}(f)$.

IV.2 INDEX OF A LINEAR OPERATOR

The notion of the index of a linear operator plays an ever-increasing role in the study of differential and integral equations. We refer the reader to Gokhberg and Krein [1] for a discussion of the development of theorems relating to an index. Recently, Atiyah and Singer [1] gave a general formula for the index of an elliptic operator on any compact oriented differentiable manifold.

IV.2.1 Definition. *The dimension of $\mathfrak{N}(T)$, written $\alpha(T)$, will be called the* ***kernel index*** *of T and the deficiency of $\mathfrak{R}(T)$ in Y, written $\beta(T)$, will be called the* ***deficiency index*** *of T. Thus $\alpha(T)$ and $\beta(T)$ will be either a nonnegative integer or ∞.*

In Example IV.1.4, $\alpha(T) = n$ and $\beta(T) = 0$.

In Example IV.1.5, $\alpha(T)$ is the number of λ_k which are 0. $\beta(T) = 0$ if $\{1/\lambda_k\}$ is a bounded sequence. $\beta(T) = \infty$ if infinitely many of the λ_k are 0.

IV.2.2 Definition. *If $\alpha(T)$ and $\beta(T)$ are not both infinite, we say that T has an index. The* ***index*** *$\kappa(T)$ is defined by*

$$\kappa(T) = \alpha(T) - \beta(T)$$

with the understanding, as in the extended real number system, that for any real number r,

$$\infty - r = \infty \qquad \text{and} \qquad r - \infty = -\infty$$

The indices of certain differential operators are determined in Chap. VI.

IV.2.3 Theorem. *Let the domain of T be dense in X. Then*

i. $\alpha(T')$ is the deficiency of $\overline{\mathfrak{R}(T)}$. In particular, if $\mathfrak{R}(T)$ is closed, then $\alpha(T') = \beta(T)$.

ii. If X and Y are complete and T is closed with closed range, then $\alpha(T) = \beta(T')$. If, in addition, T has an index, then T' has an index and $\kappa(T) = -\kappa(T')$. [We agree to let $-(-\infty) = \infty$.]

Proof of (i). By Theorems I.6.4 and II.3.7,

$$\text{(1)} \quad \dim \frac{Y}{\overline{\mathfrak{R}(T)}} = \dim \left(\frac{Y}{\overline{\mathfrak{R}(T)}} \right)' = \dim \mathfrak{R}(T)^{\perp} = \dim \mathfrak{N}(T') = \alpha(T')$$

Proof of (ii). By Theorems IV.1.2 and I.6.4,

$$\text{(2)} \quad \beta(T') = \dim \frac{X'}{\mathfrak{R}(T')} = \dim \frac{X'}{\mathfrak{N}(T)^{\perp}} = \dim \mathfrak{N}(T)' = \dim \mathfrak{N}(T) = \alpha(T)$$

Hence (1) and (2) give

$$\kappa(T) = \alpha(T) - \beta(T) = \beta(T') - \alpha(T') = -\kappa(T')$$

IV.2.4 Definition. *A closed linear operator with closed range is called* ***normally solvable.***

A bounded linear operator which has a finite index and which is defined on a Banach space is oftentimes referred to in the literature as a Fredholm operator. Fredholm [1] studied certain integral equations which gave rise to such operators. In view of the development of the theory of unbounded operators, the following definition is frequently used.

IV.2.5 Definition. *A closed linear operator which has a finite index is called a* ***Fredholm operator.***

IV.2.6 Remark. Corollary IV.1.13 shows that, in a Banach space, a Fredholm operator is normally solvable.

Examples of differential operators which are Fredholm operators are given in Chap. VI.

The following theorem was proved by Gokhberg and Krein [1] for T and B Fredholm operators.

IV.2.7 Theorem. *Let X and Y be complete and let T be normally solvable with finite kernel index; that is, $\alpha(T) < \infty$. Suppose B is a linear operator with domain a subspace of a normed linear space Z and range in X.*

i. If B is closed, then TB is closed.

ii. If B is normally solvable, then TB is normally solvable.

iii. Let the domain of T be dense in X and let Z be complete. If T and B are Fredholm operators, then TB is a Fredholm operator and

$$\kappa(TB) = \kappa(T) + \kappa(B)$$

iv. Let Z be complete and let the domains of T and B be dense in X and Z, respectively. If B is closed with finite deficiency index, that is, $\beta(B) < \infty$, then $\mathfrak{D}(TB)$ is dense in Z.

We first prove the following lemmas.

IV.2.8 Lemma. *Let M be a closed subspace having finite deficiency in X. Then*

i. For any subspace V of X there exists a finite-dimensional subspace N contained in V such that

$$\bar{V} = \bar{V} \cap M \oplus N$$

ii. If V is dense in X, then $V \cap M$ is dense in M.

Proof of (i). The dimension of $\bar{V}/\bar{V} \cap M$ does not exceed the dimension of the finite-dimensional space X/M, since the linear map η from $\bar{V}/\bar{V} \cap M$ to X/M defined by $\eta(x + \bar{V} \cap M) = x + M$ is 1-1. Hence there exists a finite-dimensional subspace W of $\bar{V}$ such that

$$\bar{V} = \bar{V} \cap M \oplus W \tag{1}$$

By Theorem II.1.14, there exists a projection P from $\bar{V}$ onto W with $P(\bar{V} \cap M) = 0$. Since PV is finite-dimensional and P is continuous, it follows that

$$W = P\bar{V} \subset \overline{PV} = PV \tag{2}$$

Let $w_1, w_2, \ldots, w_n$ be a basis for W. Then, by (2), there exist elements $v_1, v_2, \ldots, v_n$ in V such that $Pv_i = w_i$, $1 \leq i \leq n$. Since the set $\{w_i\}$ is linearly independent, the space $N = \text{sp}\,\{v_1, v_2, \ldots, v_n\}$ is an n-dimensional subspace of V and $(0) = N \cap (M \cap \bar{V})$. By (1),

$$\dim \frac{\bar{V}}{M \cap \bar{V}} = \dim W = \dim N$$

Hence it follows that $\bar{V} = M \cap \bar{V} \oplus N$.

Proof of (ii). Suppose $\bar{V} = X$. Then from (i), $X = M \oplus N$, where N is a finite-dimensional subspace of V. Therefore $V = M \cap V \oplus N$. It follows from Theorem II.1.14 that M is isomorphic to X/N under the map η, where $\eta(m) = m + N$. For clarity we write

$$M \xrightarrow{\eta} \frac{X}{N} \supset \frac{V}{N} = \frac{(M \cap V) \oplus N}{N} \xrightarrow{\eta^{-1}} M \cap V \tag{3}$$

Since V is dense in X, V/N is dense in X/N. Noting that

$$\eta(M \cap V) = \frac{V}{N}$$

it follows from (3) and the continuity of η and η^{-1} that

$$M \cap V = \eta^{-1}\eta(M \cap V)$$

is dense in M.

IV.2.9 Lemma. *Let X and Y be complete and let T be normally solvable. If M is a subspace (not necessarily closed) of X such that $M + \mathfrak{N}(T)$ is closed, then TM is closed. In particular, if M is closed and $\mathfrak{N}(T)$ is finite-dimensional, then TM is closed.*

Proof. Let T_1 be the operator T restricted to $\mathfrak{D}(T) \cap (M + \mathfrak{N}(T))$. Then T_1 is closed and $\mathfrak{N}(T_1) = \mathfrak{N}(T)$. Hence $\gamma(T_1) \geq \gamma(T) > 0$. Therefore T_1 has a closed range; that is, $TM = T_1(M + \mathfrak{N}(T))$ is closed.

Proof of Theorem IV.2.7

Proof of (i). Suppose $z_n \to z$ and $TBz_n \to y$. Since $\gamma(T)$ is positive, it follows that the sequence $\{[Bz_n]\}$ in $\mathfrak{D}(T)/\mathfrak{N}(T)$ is a Cauchy sequence. Thus there exists and $[x] \in X/\mathfrak{N}(T)$ such that $[Bz_n] \to [x]$. This implies the existence of a sequence $\{x_n\}$ in $\mathfrak{N}(T)$ such that $Bz_n + x_n \to x$. We first show that $\{x_n\}$ is bounded. Suppose $\{x_n\}$ is unbounded. Then there exists a subsequence $\{x_{n'}\}$ of $\{x_n\}$ such that

$$\|x_{n'}\| \to \infty \qquad \text{and} \qquad \frac{Bz_{n'} + x_{n'}}{\|x_{n'}\|} \to 0$$

Since $\{x_{n'}/\|x_{n'}\|\}$ is a bounded sequence in the finite-dimensional space $\mathfrak{N}(T)$, there exists a subsequence $\{x_{n''}\}$ of $\{x_{n'}\}$ and a $v \in \mathfrak{N}(T)$ such that $x_{n''}/\|x_{n''}\| \to v$. It follows that

$$\frac{Bz_{n''}}{\|x_{n''}\|} \to -v \qquad \text{and} \qquad \frac{z_{n''}}{\|x_{n''}\|} \to 0$$

Since B is closed, $v = B(0) = 0$, which cannot be, since $\|v\| = 1$. Therefore $\{x_n\}$ is a bounded sequence in $\mathfrak{N}(T)$. Consequently, there exists a subsequence $\{x_{n'}\}$ of $\{x_n\}$ and a $w \in \mathfrak{N}(T)$ such that $x_{n'} \to w$. Hence $Bz_{n'} \to x - w$. Since $z_{n'} \to z$ and B is closed, z is in $\mathfrak{D}(B)$ and

$$Bz = x - w = \lim_{n' \to \infty} Bz_{n'}$$

Since T is closed and $TBz_{n'} \to y$, Bz is in $\mathfrak{D}(T)$ and $TBz = y$, showing that TB is closed.

Proof of (ii). If BZ is closed, then $\mathcal{R}(TB) = TBZ$ is closed by Lemma IV.2.9.

Proof of (iii). The linear map η from $\mathfrak{N}(TB)/\mathfrak{N}(B)$ to $\mathcal{R}(B) \cap \mathfrak{N}(T)$, defined by $\eta[x] = Bx$, is 1-1 and surjective. Hence, for

$$N_1 = \mathcal{R}(B) \cap \mathfrak{N}(T)$$

$$\text{(1)} \qquad \alpha(TB) = \alpha(B) + n_1 \qquad n_1 = \dim N_1$$

Let N_2 be a subspace of $\mathfrak{N}(T)$ such that $\mathfrak{N}(T) = N_1 \oplus N_2$. Then

$$\text{(2)} \qquad \alpha(T) = n_1 + n_2 \qquad n_2 = \dim N_2$$

The subspaces $\mathcal{R}(B)$ and N_2 are linearly independent, since $Bx \in N_2 \subset \mathfrak{N}(T)$ implies $Bx \in N_1 \cap N_2 = (0)$.

Since $\beta(B)$ and dim N_2 are finite, $\mathcal{R}(B)$ and $\mathcal{R}(B) \oplus N_2$ are closed by Corollary IV.1.13 and Theorem I.4.12. By hypothesis, $\mathfrak{D}(T)$ is dense in X. Thus, by Lemma IV.2.8,

$$\mathcal{R}(B) \oplus N_2 \oplus N_3 = X \tag{3}$$

where N_3 is some finite-dimensional subspace of $\mathfrak{D}(T)$. Hence

$$\beta(B) = n_2 + n_3 \qquad n_3 = \dim N_3 \tag{4}$$

Now $\mathfrak{N}(T) = N_1 \oplus N_2 \subset \mathcal{R}(B) \oplus N_2$. This, together with (3), implies that T is 1-1 on all of N_3 and

$$TX = T\mathcal{R}(B) \oplus TN_3 \tag{5}$$

(In general, if $X = U \oplus V$, where $\mathfrak{N}(T) \subset U$ and V is a subspace of $\mathfrak{D}(T)$, then T is 1-1 on V and $TX = TU \oplus TV$.) It follows from (5) and the fact that T is 1-1 on all of N_3 that

$$\beta(TB) = \beta(T) + \dim TN_3 = \beta(T) + n_3 \tag{6}$$

From (1), (2), (4), and (6) we obtain

$$\kappa(TB) = \alpha(B) + n_1 - \beta(T) - n_3 = \alpha(B) + \alpha(T) - n_2 - \beta(T) - n_3$$

$$= \alpha(B) + \alpha(T) - \beta(B) - \beta(T) = \kappa(T) + \kappa(B)$$

Proof of (iv). By Corollaries IV.1.13 and IV.1.6, the 1-1 operator $\hat{B}$ induced by B has a bounded inverse on the closed subspace $\mathcal{R}(B)$. Since $\mathfrak{D}(T)$ is dense in X, $\mathfrak{D}(T) \cap \mathcal{R}(B)$ is dense in $\mathcal{R}(B)$ by Lemma IV.2.8. Thus, by the continuity of $\hat{B}^{-1}$, it follows that

$$\frac{\mathfrak{D}(B)}{\mathfrak{N}(B)} = \hat{B}^{-1}\mathcal{R}(B) = \hat{B}^{-1}\overline{\mathfrak{D}(T) \cap \mathcal{R}(B)} \subset \overline{\hat{B}^{-1}\mathfrak{D}(T) \cap \mathcal{R}(B)} \tag{7}$$

$$= \overline{\mathfrak{D}(TB)/\mathfrak{N}(B)}$$

Since $\mathfrak{D}(B)$ is dense in Z, (7) implies that $\mathfrak{D}(TB)$ is dense in Z.

Multiplying a Fredholm operator on the left by a normally solvable operator does not guarantee that the product is even a closed operator. This is shown in the following example.

IV.2.10 Example. Let T be the derivative operator in Example II.1.4. Let $B: X \to X = C([0, 1])$ be defined as the linear operator which takes each $x \in X$ into the constant function $x(0)$. B is obviously a projection. However, BT is not closed. Let $x_n(t) = e^{-nt}/n$. Then $x_n \to 0$ and BTx_n is the constant function -1, but $\lim_{n\to\infty} BTx_n = -1 \neq BT0 = 0$.

***IV.2.11* Definition.** *Suppose $X = Y$. Given a polynomial*

$$p(\lambda) = \sum_{k=0}^{n} a_k\lambda^k$$

define $p(T) = \sum_{k=0}^{n} a_kT^k$, where T^0 is the identity operator I defined on all of X. The domain of $p(T)$ is the domain of T^n.

***IV.2.12* Corollary.** *Let $X = Y$ and let X be complete. If there exists a scalar λ_0 such that $\lambda_0I - T$ is normally solvable and has finite kernel index, then for any polynomial p, the operator $p(T)$ is closed.*

Proof. Let the degree of p be n. The proof is by induction. For $n = 0$ the corollary is trivial. Assume the corollary to be valid for $n = k$ and let v be a polynomial of degree $k + 1$. We write

$$v(\lambda) = (\lambda_0 - \lambda)q(\lambda) + c$$

where q is a polynomial of degree k and c is a constant. Then

$$v(T) = (\lambda_0I - T)q(T) + cI$$

It suffices to prove that $(\lambda_0I - T)q(T)$ is closed. By the induction hypothesis, $q(T)$ is closed. Hence, by (*i*) of Theorem IV.2.7, $(\lambda_0I - T)q(T)$ is closed.

***IV.2.13* Definition.** *Let $X = Y$. The set of scalars λ for which $\lambda I - T$ is a Fredholm operator is called the* **Fredholm resolvent** *of T, denoted by $F\rho(T)$. The set of λ for which $\lambda \notin F\rho(T)$ is called the* **Fredholm spectrum** *of T, denoted by $F\sigma(T)$.*

The next theorem was proved by Rota [1], Theorem 3.2, for differential operators and later generalized by Balslev and Gamelin [1] to Fredholm operators. The proof makes use of Theorem IV.2.7.

***IV.2.14* Theorem.** *Let $X = Y$ be a Banach space over the complex numbers and let T be closed. For any polynomial p,*

i. *$p(F\sigma(T)) \subset F\sigma(p(T))$. Equality holds if $\mathfrak{D}(T)$ is dense in X.*
ii. *Suppose $\mathfrak{D}(T)$ is dense in X. If there exists a $\mu \in p(F\rho(T))$, then*

$p(T)$ is closed and has domain dense in X. Moreover

$$\kappa(\mu I - p(T)) = \sum_{i=1}^{n} \kappa(\lambda_i I - T)$$

where $\lambda_1, \lambda_2, \ldots, \lambda_n$ are the zeros of the polynomial $\mu - p(\lambda)$ counted according to their multiplicity.

Proof of (i). Given scalar ν, let $\lambda_1, \lambda_2, \ldots, \lambda_n$ be such that

$$(1) \qquad \nu - p(\lambda) = (\lambda_1 - \lambda)(\lambda_2 - \lambda) \ldots (\lambda_n - \lambda)$$

Then

$$(2) \qquad \nu I - p(T) = (\lambda_1 I - T)(\lambda_2 I - T) \ldots (\lambda_n I - T)$$

Suppose ν is in $p(F\sigma(T))$. Then, by (1), one of the λ_i is in $F\sigma(T)$, say λ_k. Since $\lambda_k I - T$ commutes with $\lambda_i I - T$, $1 \leq i \leq n$, it is obvious from (2) that

$$(3) \quad \mathfrak{N}(\lambda_k I - T) \subset \mathfrak{N}(\nu I - p(T)) \quad \text{and} \quad \mathfrak{R}(\nu I - p(T)) \subset \mathfrak{R}(\lambda_k I - T)$$

Since $\alpha(\lambda_k I - T) = \infty$ or $\beta(\lambda_k I - T) = \infty$, it follows from (3) that $\nu I - p(T)$ is not a Fredholm operator; that is, $\nu \in F\sigma(p(T))$.

Suppose $\mathfrak{D}(T)$ is dense in X. Then for $1 \leq i \leq n$, $\lambda_i I - T$ has domain dense in X. If each $\lambda_i I - T$ is a Fredholm operator, then it follows from (2) and Theorem IV.2.7 that $\nu I - p(T)$ is a Fredholm operator. Hence, if ν is in $F\sigma(p(T))$, then some $\lambda_k I - T$ is not a Fredholm operator. Thus, $F\sigma(p(T)) \subset p(F\sigma(T))$ which, when combined with what was shown, proves $F\sigma(p(T)) = p(F\sigma(T))$.

Proof of (ii). Suppose we have the representations given in (1) and (2) with μ in place of ν. It was shown in the proof of *(i)* that $\mu I - p(T)$ is a Fredholm operator if and only if each $\lambda_i I - T$ is a Fredholm operator. Thus *(ii)* follows from Theorem IV.2.7.

chapter V

PERTURBATION THEORY

One of the prime motivations for the study of perturbation theory can, perhaps, be best described by the following situation.

An operator A is given about which certain properties are to be determined; e.g., find $\kappa(A)$. If A is "complicated," one tries to express A in the form $A = T + B$, where T is a "simple" operator and B is so related to T that knowledge about properties of T is enough to gain information about the corresponding properties of A. For example, if A is a differential operator $\sum_{k=0}^{n} a_k D^k$, T might be chosen to be $a_n D^n$, with B as the remaining lower-order terms. A theory which tells us under what conditions B can be disregarded in the determination of $\kappa(A)$, for example, is very useful. Some of the numerous applications of perturbation theory appear in Chap. VI.

Most of the material presented in this chapter is based on portions of the deep results obtained by Gokhberg and Krein [1] and Kato [1]. The first paper also contains an extensive bibliography which is supplemented by Bartle in the American Mathematical Society Translations, ser. 2, vol. 13.

Throughout this chapter, B is a linear operator with domain a subspace of normed linear space X and range a subspace of normed linear space Y. As *in the previous chapters, T is a linear operator with domain in X and range in Y. The spaces are assumed complete and T is assumed closed only when specifically stated. $\mathfrak{D}(T)$ is not required to be dense in X.*

V.1 PERTURBATION BY BOUNDED OPERATORS

The following lemma, which is due to Krein, Krasnoselskii, and Milman [1], is needed throughout this chapter. The proof depends on Borsuk's antipodal-mapping theorem [1] and will be omitted. We refer the reader to Gokhberg and Krein [1], Theorem 1.1, and to Day [1].

V.1.1 Lemma. *Let M and N be subspaces of X with* $\dim M > \dim N$ *(thus* $\dim N < \infty$*). Then there exists an $m \neq 0$ in M such that*

$$\|m\| = d(m, N)$$

The lemma need not hold if $\dim M = \dim N < \infty$, as may be seen by taking X to be the plane, with M and N nonperpendicular lines through the origin.

When X is a Hilbert space, the lemma has the following easy proof. We first note that $M \cap N^\perp \neq (0)$; otherwise

$$\dim N = \dim\left(\frac{N \oplus N^\perp}{N^\perp}\right) = \dim \frac{X}{N^\perp} \geq \dim \frac{M \oplus N^\perp}{N^\perp} = \dim M$$

which contradicts the hypothesis. Take $x \neq 0$ in $M \cap N^\perp$. Then

$$\|x - n\|^2 = \|x\|^2 + \|n\|^2 \geq \|x\|^2 \qquad n \in N$$

Thus $d(x, N) = \|x\|$.

V.1.2 Theorem. *Suppose $\gamma(T) > 0$. Let B be bounded with $\mathfrak{D}(B) \supset \mathfrak{D}(T)$. If $\|B\| < \gamma(T)$, then*

i. $\alpha(T + B) \leq \alpha(T)$
ii. $\dim Y/\overline{\mathfrak{R}(T + B)} \leq \dim Y/\overline{\mathfrak{R}(T)}$

Proof of (i). For $x \neq 0$ in $\mathfrak{N}(T + B)$ and $\|B\| < \gamma = \gamma(T)$,

$$\gamma\|[x]\| \leq \|Tx\| = \|Bx\| \leq \|B\|\,\|x\| < \gamma\|x\|$$

where $[x] \in X/\mathfrak{N}(T)$. Thus $\|x\| > \|[x]\| = d(x, \mathfrak{N}(T))$, and therefore, by Lemma V.1.1, $\alpha(T + B) \leq \alpha(T)$.

Proof of (ii). Let $X_1 = \overline{\mathfrak{D}(T)}$ and let B_1 be B restricted to $\mathfrak{D}(T)$. Considering T and B_1 as operators with domains dense in X_1, the conjugates T' and B'_1 exist with domains in Y' and ranges in X'_1. By Theorem IV.1.8,

$$\gamma(T') = \gamma(T) > \|B\| \geq \|B_1\| = \|B'_1\|$$

Hence, it follows from (*i*) applied to T' and B'_1 and from Theorem IV.2.3, that

$$\dim \frac{Y}{\overline{\mathfrak{R}(T + B)}} = \dim \frac{Y}{\overline{\mathfrak{R}(T + B_1)}} = \alpha(T' + B'_1) \leq \alpha(T') = \dim \frac{Y}{\overline{\mathfrak{R}(T)}}$$

V.1.3 Corollary. *Suppose T has a bounded inverse. If B is bounded with $\|B\| < \gamma = 1/\|T^{-1}\|$, then $T + B$ has a bounded inverse. If, in addition, $\mathfrak{D}(B) \supset \mathfrak{D}(T)$, then*

$$\dim \frac{Y}{\overline{\mathfrak{R}(T)}} = \dim \frac{Y}{\overline{\mathfrak{R}(T + B)}}$$

Proof. For $x \in \mathfrak{D}(T + B)$ choose a positive integer N so that

$$\frac{\|B\|}{N} < \gamma - \|B\|$$

Now for $0 \leq k \leq N$,

$$\left\| \left(T + B - \frac{k}{N} B \right) x \right\| \geq \|Tx\| - \left\| \left(1 - \frac{k}{N} \right) Bx \right\| \geq (\gamma - \|B\|) \|x\|$$

Thus $T + B$ has a bounded inverse and

$$\gamma\left(T + B - \frac{k}{N} B\right) \geq \gamma - \|B\| \qquad 0 \leq k \leq N \tag{1}$$

Since $\|(1/N)B\| < \gamma - \|B\|$, Theorem V.1.2, together with (1), implies that for $0 \leq k \leq N$ and $\bar{\beta}(T) = \dim Y/\overline{\mathfrak{R}(T)}$,

$$\bar{\beta}\left(T + B - \frac{k+1}{N} B\right) = \bar{\beta}\left(T + B - \frac{k}{N} B - \frac{1}{N} B\right) \leq \bar{\beta}\left(T + B - \frac{k}{N} B\right) \tag{2}$$

It follows from (2) and Theorem V.1.2 that

$$\bar{\beta}(T) \leq \bar{\beta}(T + B) \leq \bar{\beta}(T)$$

which proves the corollary.

As a consequence of the corollary we obtain the following result.

V.1.4 Corollary. *Let B be bounded with $\mathfrak{D}(B) \supset \mathfrak{D}(T)$. Then the B-resolvent of $T = \{\lambda \mid T - \lambda B$ has a dense range and a bounded inverse$\}$ is an open set.*

V.1.5 Lemma. *Suppose that T_1 is a linear extension of T such that $\infty > n = \dim \mathfrak{D}(T_1)/\mathfrak{D}(T)$.*

i. If T is closed, then T_1 is closed.
ii. If T has a closed range, then T_1 has a closed range.
iii. If T has an index, then $\kappa(T_1) = \kappa(T) + n$.

Proof of (i). By hypothesis, $\mathfrak{D}(T_1) = \mathfrak{D}(T) \oplus N$, where N is a finite-dimensional subspace. Hence, $G(T_1) = G(T) + Z$, where $G(T)$ and $G(T_1)$ are the graphs of T and T_1, respectively, and $Z = \{(n, T_1 n) \mid n \in N\}$. Thus, if $G(T)$ is closed, then $G(T_1)$ is closed, since Z is finite-dimensional.

Proof of (ii). If $\mathfrak{R}(T)$ is closed, then $\mathfrak{R}(T_1)$ is closed, since

$$\mathfrak{R}(T_1) = \mathfrak{R}(T) + T_1 N$$

and $T_1 N$ is finite-dimensional.

Proof of (iii). It is easy to see that it suffices to prove (*iii*) for the case when $n = 1$. Suppose that $\mathfrak{D}(T_1) = \mathfrak{D}(T) \oplus \text{sp}\,\{x\}$, for some $x \in \mathfrak{D}(T_1)$. Then $T_1 X = TX \oplus V$, where $V = \text{sp}\,\{T_1 x\}$ when $T_1 x \notin \mathfrak{R}(T)$ or $V = (0)$ when $T_1 x \in \mathfrak{R}(T)$.

If $T_1 x \notin \mathfrak{R}(T)$, then it is easy to verify that $\beta(T) = \beta(T_1) + 1$ and that $\mathfrak{N}(T) = \mathfrak{N}(T_1)$. Therefore, $\kappa(T_1) = \kappa(T) + 1$.

If $T_1 x \in \mathfrak{R}(T)$, then $\mathfrak{R}(T_1) = \mathfrak{R}(T)$ and there exists a $z \in \mathfrak{D}(T)$ such that $Tz = T_1 x$. Hence, $\mathfrak{N}(T_1) = \mathfrak{N}(T) \oplus \text{sp}\,\{x - z\}$. Thus $\alpha(T_1) = \alpha(T) + 1$ and $\kappa(T_1) = \kappa(T) + 1$.

The following theorem is due to Kato. Using Corollary V.1.3, we give a simplified proof of (*iii*).

V.1.6 Theorem. *Let X and Y be complete and let $\mathfrak{D}(T) \subset \mathfrak{D}(B)$. Suppose that T is normally solvable and has an index. If $\|B\| < \gamma(T)$, then*

i. $T + B$ *is normally solvable.*
ii. $\alpha(T + B) \leq \alpha(T); \beta(T + B) \leq \beta(T)$
iii. $\kappa(T + B) = \kappa(T)$

Proofs of (i) and (ii). By Theorem V.1.2, $\alpha(T + B) \leq \alpha(T)$. It is easy to verify that $T + B$ is closed. Suppose $\alpha(T) < \infty$. If $\mathcal{R}(T + B)$ is not closed, there exists, by Corollary III.1.10, a closed infinite-dimensional subspace $M \subset \mathfrak{D}(T + B)$ such that for $\gamma = \gamma(T)$ and all $x \in M$,

$$\|(T + B)x\| < (\gamma - \|B\|)\|x\|$$

Hence, for each $x \neq 0$ in M

$$\gamma d(x, \mathfrak{N}(T)) \leq \|Tx\| \leq \|Bx\| + \|(T + B)x\| < (\|B\| + \gamma - \|B\|)\|x\|$$

Therefore $d(x, \mathfrak{N}(T)) < \|x\|, 0 \neq x \in M$, whence $\dim M \leq \alpha(T) < \infty$ by Lemma V.1.1. But this contradicts $\dim M = \infty$. Hence $T + B$ is normally solvable.

Suppose $\beta(T) < \infty$. Let X_1 and B_1 be defined as in the proof of *(ii)* of Theorem V.1.2, where it was shown that $\|B_1\| < \gamma(T')$. Since $\alpha(T') = \beta(T) < \infty$ by Theorem IV.2.3, we may apply the result just proved to T' and B_1' to conclude that $T' + B_1'$ is normally solvable and

$$\beta(T) = \alpha(T') \geq \alpha(T' + B_1') \tag{1}$$

Since $T' + B_1'$ is normally solvable and $T' + B_1' = (T + B_1)'$, Theorems IV.1.2 and IV.2.3 show that $\mathcal{R}(T + B) = \mathcal{R}(T + B_1)$ is closed and

$$\alpha(T' + B_1') = \beta(T + B) \tag{2}$$

Thus $\beta(T) \geq \beta(T + B)$ by (1) and (2).

Proof of (iii). Let I denote the closed interval $[0, 1]$ and let Z be the set of integers together with the "ideal" elements ∞ and $-\infty$. Define $\varphi: I \to Z$ by $\varphi(\lambda) = \kappa(T + \lambda B)$. Let I have the usual topology and let Z have the discrete topology; i.e., points are open sets. To prove *(iii)*, it suffices to show that φ is continuous. Indeed, if φ is continuous, then $\varphi(I)$ is a connected set which therefore consists of only one point. In particular, $\kappa(T) = \varphi(0) = \varphi(1) = \kappa(T + B)$.

In order to prove the continuity of φ, we first show that

$$\kappa(T + B) = \kappa(T)$$

for $\|B\|$ sufficiently small. Suppose $\alpha(T) < \infty$. Then there exists a closed subspace M of X such that

$$X = M \oplus \mathfrak{N}(T)$$

Let T_M be the restriction of T to $M \cap \mathfrak{D}(T)$. Then T_M has a bounded inverse since it is closed, 1-1, and $\mathfrak{R}(T_M) = \mathfrak{R}(T)$. Thus by (*i*), (*ii*), and Corollary V.1.3

$$(1) \qquad \alpha(T_M + B) = \alpha(T_M) = 0 \qquad \beta(T_M + B) = \beta(T)$$

provided $\|B\| < \gamma(T_M)$. Now

$$\mathfrak{D}(T + B) = \mathfrak{D}(T) = \mathfrak{D}(T) \cap M \oplus \mathfrak{N}(T)$$

$$\mathfrak{D}(T_M + B) = \mathfrak{D}(T_M) = \mathfrak{D}(T) \cap M$$

Hence (1) and Lemma V.1.5 imply that for $\|B\| < \gamma(T_M)$,

$$(2) \qquad \kappa(T) = \kappa(T_M) + \alpha(T) = \kappa(T_M + B) + \alpha(T) = \kappa(T + B)$$

Suppose $\beta(T) < \infty$. Then, as in the proof of (*ii*), we apply (2) to T' and B_1', and thereby obtain

$$(3) \qquad \kappa(T + B) = \kappa(T + B_1) = -\kappa(T' + B_1') = -\kappa(T') = \kappa(T)$$

for $\|B\|$ sufficiently small.

Given $\lambda_0 \in [0, 1]$, $T + \lambda_0 B$ is normally solvable and has an index by (*i*) and (*ii*). Thus, by the results obtained above, it follows that

$$\varphi(\lambda) = \kappa(T + \lambda_0 B + (\lambda - \lambda_0)B) = \kappa(T + \lambda_0 B) = \varphi(\lambda_0)$$

for λ sufficiently close to λ_0. Hence φ is continuous. This completes the proof of (*iii*).

V.1.7 Corollary. *Let X, Y, T, and B satisfy the hypotheses in Theorem* V.1.6. *There exists a number $\rho > 0$ such that $\alpha(T + \lambda B)$ and $\beta(T + \lambda B)$ are constant in the annulus $0 < |\lambda| < \rho$.*

Proof. The proof of the corollary is a modification of Lemma 8.1 in Gokhberg and Krein [1], where B was taken to be the identity operator.

We first assume $\alpha(T) < \infty$. For $x \in \mathfrak{N}(T + \lambda B)$ and $\lambda \neq 0$,

$$Tx = -\lambda Bx$$

whence

$$Bx \in \mathfrak{R}(T) = R_1 \qquad \text{and} \qquad x \in B^{-1}R_1 = D_1$$

Thus

$$-\lambda Bx = Tx \in TD_1 = R_2 \qquad \text{and} \qquad x \in B^{-1}R_2 = D_2$$

It follows that

$$\mathfrak{N}(T + \lambda B) \subset \bigcap_{k=1}^{\infty} D_k \tag{1}$$

where

$$D_k = B^{-1}R_k \qquad R_1 = \mathfrak{R}(T)$$

and

$$R_{k+1} = TD_k$$

It is easy to see that

$$R_1 \supset R_2 \supset \cdots \qquad \text{and} \qquad D_1 \supset D_2 \supset \cdots$$

We show by induction that R_n and D_n are closed subspaces of Y and X, respectively. R_1 is closed by hypothesis, and D_1 is closed since B is continuous. Suppose R_k and D_k are closed. By Lemma IV.2.9, $R_{k+1} = TD_k$ is closed and therefore $D_{k+1} = B^{-1}R_{k+1}$ is closed since B is continuous. Hence D_n and B_n are closed by induction. Define

$$X_1 = \bigcap_{k=1}^{\infty} D_k \qquad \text{and} \qquad Y_1 = \bigcap_{k=1}^{\infty} R_k$$

It follows from the definition of D_k and R_k that $TX_1 \subset Y_1$ and $BX_1 \subset Y_1$. Let T_1 and B_1 be the operators T and B, respectively, restricted to $\mathfrak{D}(T) \cap X_1$ with ranges in Y_1. Since T is a closed operator and X_1 is closed, T_1 is also closed. We show that $\mathfrak{R}(T_1) = Y_1$. Let y be an element in $Y_1 = \bigcap_{n=1}^{\infty} TD_n$. Then for each $n \geq 1$ there exists an $x_n \in D_n$ such that $Tx_n = y$. Since $\mathfrak{N}(T)$ is finite-dimensional and $D_n \supset D_{n+1}$, there exists an integer k_0 such that

$$\mathfrak{N}(T) \cap D_{k_0} = \mathfrak{N}(T) \cap D_k \qquad \text{for } k \geq k_0$$

From the way the sequence $\{x_k\}$ was chosen, together with the fact that $D_k \subset D_{k_0}$, $k \geq k_0$, it follows that

$$x_k - x_{k_0} \in \mathfrak{N}(T) \cap D_{k_0} = \mathfrak{N}(T) \cap D_k \subset D_k \qquad k \geq k_0$$

Hence $x_{k_0} \in \bigcap_{k \geq k_0} D_k = X_1$ and $Tx_{k_0} = y$, which shows that T_1 is surjective. Thus there exists, by Theorem V.1.6, a number $\rho > 0$ such that for $|\lambda| < \rho$,

$$\kappa(T + \lambda B) = \kappa(T) \tag{2}$$

$$\beta(T_1 + \lambda B_1) = \beta(T_1) = 0 \quad \text{and} \quad \alpha(T_1 + \lambda B_1) = \kappa(T_1) = \alpha(T_1) \tag{3}$$

Now, (1) implies that $\mathfrak{N}(T + \lambda B) = \mathfrak{N}(T_1 + \lambda B_1)$ for $\lambda \neq 0$. In particular,

$$\text{(4)} \qquad \alpha(T + \lambda B) = \alpha(T_1 + \lambda B_1) \qquad \lambda \neq 0$$

It is clear from (2), (3) and (4) that $\alpha(T + \lambda B)$ and $\beta(T + \lambda B)$ are constant in the annulus $0 < |\lambda| < \rho$ under the assumption $\alpha(T) < \infty$. If $\alpha(T) = \infty$, then $\beta(T) < \infty$ by hypothesis. As in the proof of Theorem V.1.6, we apply the result just proved to the appropriate conjugate operators in order to prove the corollary when $\alpha(T) = \infty$.

V.1.8 Theorem. *Let X and Y be complete and let B be continuous with* $\mathfrak{D}(T) \subset \mathfrak{D}(B)$. *Define U to be the set of* λ *for which* $T + \lambda B$ *is normally solvable and has an index. Then*

i. U is an open set.
ii. If C is a component of U, that is, a largest connected subset of U, then on C, with the possible exception of isolated points, $\alpha(T + \lambda B)$ *and* $\beta(T + \lambda B)$ *have constant values* n_1 *and* n_2, *respectively. At the isolated points,*

$$\alpha(T + \lambda B) > n_1 \qquad \textit{and} \qquad \beta(T + \lambda B) > n_2$$

Proof of (i). For $\lambda \in U$, apply Theorem V.1.6 to $T + \lambda B$ in place of T.

Proof of (ii). The component C is open, since any component of an open set in the space of scalars is open. Let $\alpha(\lambda_0) = n_1$ be the smallest integer which is attained by $\alpha(\lambda) = \alpha(T + \lambda B)$ on C. Suppose $\alpha(\lambda') \neq n_1$. Owing to the connectivity of C, there exists an arc Γ lying in C with endpoints λ_0 and λ'. It follows from Corollary V.1.7 and the fact that C is open, that about each $\mu \in \Gamma$ there exists an open ball $S(\mu)$ contained in C such that $\alpha(\lambda)$ is constant on the set $S(\mu)$ with the point μ deleted.

Since Γ is compact and connected, there exist points

$$\lambda_1, \lambda_2, \ldots, \lambda_n = \lambda'$$

on Γ such that

$$\text{(1)} \quad S(\lambda_0), S(\lambda), \ldots, S(\lambda_n) \text{ cover } \Gamma \text{ and } S(\lambda_i) \cap S(\lambda_{i+1}) \neq \phi \qquad 0 \leq i \leq n - 1$$

We assert that $\alpha(\lambda) = \alpha(\lambda_0)$ on all of $S(\lambda_0)$. Indeed, it follows from Theorem V.1.6 that $\alpha(\lambda) \leq \alpha(\lambda_0)$ for λ sufficiently close to λ_0. Therefore, since $\alpha(\lambda_0)$ is the minimum of $\alpha(\lambda)$ on C, $\alpha(\lambda) = \alpha(\lambda_0)$ for λ sufficiently close to λ_0. Since $\alpha(\lambda)$ is constant for all $\lambda \neq \lambda_0$ in $S(\lambda_0)$, this constant must be $\alpha(\lambda_0)$. Now $\alpha(\lambda)$ is constant on the set $S(\lambda_i)$ with the point λ_i

deleted, $1 \leq i \leq n$. Hence, it follows from (1) and the observation $\alpha(\lambda) = \alpha(\lambda_0)$ for all $\lambda \in S(\lambda_0)$, that $\alpha(\lambda) = \alpha(\lambda_0)$ for all $\lambda \neq \lambda'$ in $S(\lambda')$ and $\alpha(\lambda') > n_1$. The result just obtained can be applied to $T' + \lambda B_1'$, as in Theorem V.1.6, in order to prove the analogous results for

$$\beta(T + \lambda B) = \alpha(T' + \lambda B_1')$$

V.1.9 Remarks. The following examples show that Theorem V.1.6 does not hold in general even though $\|B\| = \gamma(T)$.

If we take T to be the identity operator mapping an infinite-dimensional Banach space onto itself, then

$$\gamma(T) = 1 = \|T\| \qquad \text{and} \qquad \alpha(T) = \beta(T) = 0$$

However, T perturbed by $-T$ is the zero operator which has ∞ as its kernel index and deficiency index.

Less trivially, in Example IV.1.5, choose $\lambda_1, \lambda_2, \ldots$ to be a strictly decreasing sequence of numbers converging to a positive number r. Then $\alpha(T) = \beta(T) = 0$, $\gamma(T) = r$, but $\gamma(T - rI) = 0$. Hence, $T - rI$ does not have a closed range, and therefore $\beta(T - rI) = \infty$ by Corollary IV.1.6.

V.2 PERTURBATION BY STRICTLY SINGULAR OPERATORS

If B is strictly singular with no restriction on its norm, we obtain the following important stability theorem due to Kato.

V.2.1 Theorem. *Let X and Y be complete and let T be normally solvable with $\alpha(T) < \infty$. If B is strictly singular and $\mathfrak{D}(T) \subset \mathfrak{D}(B)$, then*

i. *$T + B$ is normally solvable.*
ii. *$\kappa(T + B) = \kappa(T)$*
iii. *$\alpha(T + \lambda B)$ and $\beta(T + \lambda B)$ have constant values n_1 and n_2, respectively, except perhaps for isolated points. At the isolated points,*

$$\infty > \alpha(T + \lambda B) > n_1 \qquad \text{and} \qquad \beta(T + \lambda B) > n_2$$

Proof of (i). Since $\alpha(T) < \infty$, there exists a closed subspace M of X such that $X = M \oplus \mathfrak{N}(T)$. Let T_M be the operator T restricted to $M \cap \mathfrak{D}(T)$. Then T_M is closed with $\mathfrak{R}(T_M) = \mathfrak{R}(T)$. Suppose that $T + B$ does not have a closed range. Then it follows from Lemma V.1.5 that the closed operator $T_M + B$ does not have a closed range. Hence there exists, by Corollary III.1.10, a closed infinite-dimensional subspace

M_0 contained in $\mathfrak{D}(T_M) = \mathfrak{D}(T_M + B)$ such that

$$\|(T_M + B)x\| < \frac{\gamma(T_M)}{2} \|x\| \qquad x \in M_0$$

Thus, since T_M is 1-1, it follows that for all x in M_0,

$$\|Bx\| \geq \|T_M x\| - \|(T_M + B)x\| \geq \frac{\gamma(T_M)}{2} \|x\|$$

which shows that B has a bounded inverse on the infinite-dimensional space M_0. This, however, contradicts the hypothesis that B is strictly singular. We now prove $\alpha(T + B) < \infty$. There exists a closed subspace N_1 such that

$$\mathfrak{N}(T + B) = \mathfrak{N}(T + B) \cap \mathfrak{N}(T) \oplus N_1 \tag{1}$$

Let T_1 be the operator T restricted to N_1. By Lemma IV.2.9, $\mathfrak{R}(T_1) = TN_1$ is closed. Moreover, T_1 is 1-1. Hence its inverse is bounded. Since $B = -T_1$ on N_1 and B is strictly singular, N_1 must be finite-dimensional. Therefore (1) implies that $\mathfrak{N}(T + B)$ is finite-dimensional.

Proof of (ii). We have shown above that for all λ, $T + \lambda B$ is normally solvable and $\alpha(T + \lambda B) < \infty$. Thus, applying Theorem V.1.6 to $T + \lambda B$ in place of T, it follows that $\varphi(\lambda) = \kappa(T + \lambda B)$ is continuous from $[0, 1]$ into Z, where $[0, 1]$ has the usual topology and Z is the topological space consisting of the integers and $-\infty$ with the discrete topology. Hence φ is a constant function. In particular,

$$\kappa(T) = \varphi(0) = \varphi(1) = \kappa(\mathrm{T} + B)$$

Proof of (iii). Since $T + \lambda B$ is normally solvable for all λ, *(iii)* is an immediate consequence of Theorem V.1.8.

V.2.2 **Corollary.** *Let X and Y be complete with $\mathfrak{D}(T)$ dense in X. Let T be normally solvable and have an index. Suppose $\mathfrak{D}(T) \subset \mathfrak{D}(B)$ and both B and B' are strictly singular; for example, B is compact. Then*

i. *$T + B$ is normally solvable.*
ii. $\kappa(T + B) = \kappa(T)$
iii. *$\alpha(T + \lambda B)$ and $\beta(T + \lambda B)$ have constant values n_1 and n_2, respectively, except perhaps for isolated points. At the isolated points,*

$$\alpha(T + \lambda B) > n_1 \qquad \textit{and} \qquad \beta(T + \lambda B) > n_2$$

Proof. Apply Theorem V.2.1 to T' and B' when $\beta(T) < \infty$.

***V.2.3* Corollary.** *Let $X = Y$ be complete. Suppose there exists a scalar λ_0 such that $(\lambda_0 I - T)^{-1}$ is strictly singular on all of Y. Then for $B \epsilon [X]$, $T + B$ is a Fredholm operator with zero index.*

Proof. We write

$$T + B = [-I + (\lambda_0 I + B)(\lambda_0 I - T)^{-1}](\lambda_0 I - T) \tag{1}$$

Since $(\lambda_0 I - T)^{-1}$ is assumed to be strictly singular on X, we have from Theorem V.2.1 that

$$\kappa(-I + (\lambda_0 I + B)(\lambda_0 I - T)^{-1} = \kappa(I) = 0 \tag{2}$$

Furthermore

$$\kappa(\lambda_0 I - T) = 0 \tag{3}$$

Thus (1), (2), (3), and Theorem IV.2.7 imply $\kappa(T + B) = 0$.

***V.2.4* Corollary.** *Let $X = Y$ be any of the spaces $\mathfrak{L}_1(S, \Sigma, \mu)$, $\mathfrak{L}_\infty(S, \Sigma, \mu)$, $C(S)$, $B(S)$, or their conjugates. Suppose there exists a scalar λ_0 such that $(\lambda_0 I - T)^{-1}$ is weakly compact on all of X. Then for $B \epsilon [X]$, $T + B$ is a Fredholm operator with index zero.*

Proof. Theorem III.3.5 and Corollary V.2.3.

Goldman [1] proved that to any normally solvable operator T which does not have an index, it is always possible to add a compact operator $B \epsilon [X, Y]$ such that for all $\lambda \neq 0$, $T + \lambda B$ does not have a closed range. To show this, we present a similar proof given by Whitley, which does not depend on a theorem of Mackey's used by Goldman.

***V.2.5* Lemma.** *The conjugate space of a separable normed linear space contains a countable total subset.*

Proof. Let $\{x_k\}$ be a countable set dense in normed linear space X. For each k there exists an $x'_k \epsilon X'$ such that $\|x'_k\| = 1$ and $x'_k x_k = \|x_k\|$. We show that $\{x'_k\}$ is total. Assume there exists an $x \neq 0$ in X such that $x'_k x = 0$ for each k. Since $\{x_k\}$ is dense in X, there is an x_N such that $\|x - x_N\| < \|x\|/2$. Hence

$$\|x_N\| \geq \|x\| - \|x - x_N\| > \frac{\|x\|}{2}$$

But

$$\|x_N\| = x'_N x_N - x'_N x \leq \|x_N - x\| < \frac{\|x\|}{2}$$

which is a contradiction. Thus $\{x'_k\}$ is total.

V.2.6 Theorem. *Let X and Y be complete and let T be normally solvable with no index. Then there exists a compact operator $B \in [X, Y]$ such that $T + \lambda B$ does not have a closed range for all $\lambda \neq 0$. If $\mathfrak{N}(T)$ is separable, then $T + \lambda B$* is 1-1.

Proof. Since $\dim \mathfrak{R}(T)^{\perp} = \dim (Y/\mathfrak{R}(T))' = \beta(T) = \infty$, there is an infinite linearly independent set $\{y_1', y_2', \ldots\} \subset \mathfrak{R}(T)^{\perp}$. Choose $y_1 \in Y$ such that $y_1'y_1 \neq 0$. For $k = 1, 2, \ldots$ let y_{k+1} be an element in $\bigcap_{i=1}^{k} \mathfrak{N}(y_i') = {}^{\perp}\text{sp}\,\{y_1', \ldots, y_k'\}$ but not in $\mathfrak{N}(y_{k+1}')$. The existence of y_{k+1} is assured by Remark II.3.6 and the linear independence of $\{y_i'\}$. Thus

$$\begin{aligned} y_j'\mathfrak{R}(T) &= 0 \qquad 1 \leq j \\ y_j'y_i &= 0 \qquad 1 \leq j < i \\ y_j'y_j &\neq 0 \end{aligned} \tag{1}$$

Let $\{x_1, x_2, \ldots\}$ be an infinite linearly independent set in $\mathfrak{N}(T)$. Since $X_1 = \overline{\text{sp}}\,\{x_i\}$ is separable, there is, by Lemma V.2.5, a countable total set $\{v_i\} \subset X_1'$. Obviously we may assume no $v_i = 0$. Let $x_i' \in X'$ be an extension of v_i. Define $B \in [X, Y]$ by

$$Bx = \frac{\sum_{i=1}^{\infty} x_i'(x)y_i}{2^i\|x_i'\|\,\|y_i\|} \tag{2}$$

The completeness of Y, together with the absolute convergence of the series, implies that Bx exists, $B \in [X, Y]$, and B is the limit in $[X, Y]$ of the operators B_n defined by

$$B_nx = \frac{\sum_{i=1}^{n} x_i'(x)y_i}{2^i\|x_i'\|\,\|y_i\|}$$

Since B_n is bounded with finite-dimensional range, B_n and, in turn, B are compact. Now $\mathfrak{R}(B) \cap \mathfrak{R}(T) = (1)$. Indeed, if Bx is in $\mathfrak{R}(T)$, then by (1) and (2), $0 = y_1'Bx = x_1'(x)y_1'y_1$, whence $x_1'x = 0$. Thus

$$0 = y_2'Bx = x_2'(x)y_2'y_2 \qquad \text{whence} \qquad x_2'x = 0$$

Continuing in this manner, we see that $0 = x_i'x$, $1 \leq i$, and consequently $Bx = 0$. In particular, B is 1-1 on $X_1 = \overline{\text{sp}}\,\{x_i\} \subset \mathfrak{N}(T)$, for $Bx = 0 \in \mathfrak{R}(T)$ implies $x_i'x = 0$, $1 \leq i$ and therefore $x = 0$, since $\{x_i'\}$ is total on X_1.

Suppose $T + \lambda B$ has a closed range for some $\lambda \neq 0$. Then

$$\|(T + \lambda B)x\| \geq \gamma\, d(x, \mathfrak{N}(T + \lambda B)) \qquad \gamma = \gamma(T + \lambda B) > 0 \tag{3}$$

Now $\mathfrak{N}(T + \lambda B) = \mathfrak{N}(B) \cap \mathfrak{N}(T)$, since $(T + \lambda B)x = 0$ implies $Bx \in \mathfrak{R}(T) \cap \mathfrak{R}(B) = (0)$ and $Tx = 0$. Hence, defining B_1 to be the restriction of B to $\mathfrak{N}(T)$, $\mathfrak{N}(B_1) = \mathfrak{N}(T) \cap \mathfrak{N}(B) = \mathfrak{N}(T + \lambda B)$. Thus, for $x \in \mathfrak{N}(T)$, we know from (3) that

$$\|B_1 x\| = \frac{1}{|\lambda|} \|(T + \lambda B_1)x\| \geq \frac{\gamma}{|\lambda|} d(x, \mathfrak{N}(B_1))$$

Hence B_1 has a closed range which must be finite-dimensional, since B_1 is compact. But this is impossible, since $\dim X_1 = \infty$ and B is 1-1 on $X_1 \subset \mathfrak{N}(T)$. Hence $\mathfrak{R}(T + \lambda B)$ is not closed.

If $\mathfrak{N}(T)$ is separable, then $\{x_i\}$ may be chosen so that $X_1 = \mathfrak{N}(T)$, in which case $\mathfrak{N}(T + \lambda B) = \mathfrak{N}(B) \cap X_1 = (0)$.

V.3 PERTURBATION BY UNBOUNDED OPERATORS

Sz.-Nagy [1] observed that certain perturbation theorems remain valid if the perturbing operator B is not necessarily bounded but satisfies some less restrictive conditions. The conditions are that B be bounded or compact on $\mathfrak{D}(T)$ under certain renormings of $\mathfrak{D}(T)$.

V.3.1 Definition. *Let the T-norm* $\| \ \|_T$ *be defined on* $\mathfrak{D}(T)$ *by*

$$\|x\|_T = \|x\| + \|Tx\|$$

B is called ***T-bounded*** *if* $\mathfrak{D}(T) \subset \mathfrak{D}(B)$ *and B is bounded on* $\mathfrak{D}(T)$ *with respect to the T-norm; i.e., there exists a number M such that for all* $x \in \mathfrak{D}(T)$,

$$\|Bx\| \leq M(\|x\| + \|Tx\|)$$

B is called ***T-strictly singular (T-compact)*** *if* $\mathfrak{D}(T) \subset \mathfrak{D}(B)$ *and B is strictly singular (compact) on* $\mathfrak{D}(T)$ *with respect to the T-norm.*

Obviously, if B is bounded and $\mathfrak{D}(T) \subset \mathfrak{D}(B)$, then B is T-bounded. Similarly, B strictly singular (compact) implies that B is T-strictly singular (T-compact).

V.3.2 Remark. It is very easy to see that if X and Y are Banach spaces and T is closed, then $\mathfrak{D}(T)$ with the T-norm is a Banach space.

V.3.3 Theorem. *Let X and Y be complete and let T be closed. If* $\mathfrak{D}(T) \subset \mathfrak{D}(B)$ *and B is closable, then B is T-bounded.*

Proof. Let D_T be $\mathfrak{D}(T)$ with the T-norm and let $\bar{B}$ be a closed linear extension of B. Then $\bar{B}$ is closed and therefore continuous on Banach space D_T. Hence B is also continuous on D_T.

V.3.4 Corollary. *Let X and Y be complete. Suppose there exists a scalar λ_0 such that $\lambda_0 I - T$ is normally solvable and has finite kernel index. Then for any polynomial p of degree n, T^k is $p(T)$-bounded for $0 \leq k \leq n$.*

Proof. Corollary IV.2.12 and Theorem V.3.3.

The method of generalizing Theorem V.1.6 stems from the following result due to Sz.-Nagy [1].

V.3.5 Lemma. *Let Y be complete and let T be closed. Suppose $\mathfrak{D}(T) \subset \mathfrak{D}(B)$. If there exist numbers a and b such that*

$$b < 1 \qquad \text{and} \qquad \|Bx\| \leq a\|x\| + b\|Tx\|$$

for all $x \in \mathfrak{D}(T)$, then $T + B$ is closed.

Proof. It follows from the hypotheses that for all $x \in \mathfrak{D}(T)$,

$$\|(T + B)x\| \leq \|Tx\| + \|Bx\| \leq a\|x\| + (1 + b)\|Tx\| \tag{1}$$

and

$$\|(T + B)x\| \geq \|Tx\| - \|Bx\| \geq -a\|x\| + (1 - b)\|Tx\|$$

or

$$\|Tx\| \leq (1 - b)^{-1}(a\|x\| + \|(T + B)x\|) \tag{2}$$

Suppose $x_n \to x$ and $(T + B)x_n \to y$. Then (2) shows that $\{Tx_n\}$ is a Cauchy sequence. Hence there exists a $y \in Y$ such that $Tx_n \to y$. Since T is closed, $x \in \mathfrak{D}(T) = \mathfrak{D}(T + B)$ and $Tx_n \to Tx$. Thus, by (1),

$$\|(T + B)(x_n - x)\| \leq a\|x_n - x\| + (1 + b)\|Tx_n - Tx\| \to 0$$

Hence $(T + B)x_n \to (T + B)x$, and the lemma is proved.

V.3.6 Theorem. *Theorem* V.1.6 *holds if we replace the requirement $\|B\| < \gamma(T)$ by*

$$\|Bx\| \leq a\|x\| + b\|Tx\| \qquad x \in \mathfrak{D}(T)$$

where a and b are nonnegative numbers such that $a + b\gamma(T) < \gamma(T)$.

Proof. By Lemma V.3.5, $T + B$ is closed. We first assume that $a > 0$. Let T_1 and B_1 be the operators T and B, respectively, considered as mappings from Banach space D_1 into Y, where D_1 is $\mathfrak{D}(T)$ with norm $\| \ \|_1$ defined by

$$\|x\|_1 = a\|x\| + b\|Tx\|$$

The idea of the proof is to apply Theorem V.1.6 to T_1 and B_1 in order to obtain the assertions about $T + B$. We therefore want to know how $\gamma(T_1)$ compares with $\gamma(T)$. This can easily be obtained in the following manner for $a > 0$ and $b > 0$. Given x in D_1,

$$d(x, \mathfrak{N}(T_1)) = \inf_{n \in \mathfrak{N}(T)} \|x - n\|_1 = a \inf_{n \in \mathfrak{N}(T)} \|x - n\| + b\|Tx\|$$
$$= a\|[x]\| + b\|Tx\|$$

where $[x] \in \mathfrak{D}(T)/\mathfrak{N}(T)$. Thus

$$\gamma(T_1) = \inf_{x \in D_1} \frac{\|Tx\|}{d(x, \mathfrak{N}(T_1))} = \inf_{x \in D_1} \frac{\|Tx\|}{a\|[x]\| + b\|Tx\|} = \inf_{x \in D_1} \frac{\|Tx\|/\|[x]\|}{a + b\|Tx\|/\|[x]\|}$$
$$= \frac{\gamma(T)}{a + b\gamma(T)} > 1$$

Since $\|B_1\| \leq 1 < \gamma(T_1)$ and T_1 is bounded on the Banach space D_1, it follows that the conclusions of Theorem V.1.6 hold for T_1 and B_1 in place of T and B, respectively. However, $\mathfrak{R}(T_1 + B_1) = \mathfrak{R}(T + B)$, and $\mathfrak{N}(T_1 + B_1) = \mathfrak{N}(T + B)$. Thus Theorem V.1.6 holds for T and B, where $a > 0$.

If $a = 0$, choose $\varepsilon > 0$ so that $\varepsilon + b\gamma(T) < \gamma(T)$. Then $\|Bx\| \leq \varepsilon\|x\| + b\|Tx\|$, and the theorem is proved by what was just shown.

V.3.7 Theorem. *Theorem* V.2.1 *holds if B is T-strictly singular.*

Proof. Let D_1, T_1, and B_1 be as in the proof of the preceding theorem with $a = b = 1$, and apply Theorem V.2.1. To see that $T + B$ is closed, let E be the identity map from $\mathfrak{D}(T)$ onto D_1. Then E is closed and $T + B = (T_1 + B_1)E$. Thus $T + B$ is closed by Theorem IV.2.7.

V.3.8 Corollary. *Suppose T and B are closed and B is T-compact. Then*

i. Given $\varepsilon > 0$, there exists a constant $K = K(\varepsilon)$ such that for all $f \in \mathfrak{D}(T)$,

$$\|Bf\| \leq K\|f\| + \varepsilon\|Tf\|$$

ii. If Y is complete, then $T + B$ is closed.

iii. If X and Y are complete, then T is normally solvable with finite kernel index if and only if $T + B$ is normally solvable with finite kernel index. In this case, $\kappa(T) = \kappa(T + B)$.

Proof of (i). Assume (*i*) is false. Then for some $\varepsilon > 0$ and each positive integer n, there exists an $f_n \in \mathfrak{D}(T)$ such that

$$\|Bf_n\| > n\|f_n\| + \varepsilon\|Tf_n\| \tag{1}$$

Let $g_n = f_n/\|f_n\|_T$, where $\| \quad \|_T$ is the T-norm. Then from (1),

$$\|Bg_n\| > n\|g_n\| + \varepsilon\|Tg_n\| \tag{2}$$

Now, $\|g_n\|_T = 1$ and B is T-compact. Hence there exists a subsequence $\{v_n\}$ of $\{g_n\}$ such that $\{Bv_n\}$ converges in Y. Since $\|v_n\| < \|Bv_n\|/n$ and $\{\|Bv_n\|\}$ is bounded, $v_n \to 0$ in X. From the assumption that B is closed it follows that $Bv_n \to 0$. Thus $Tv_n \to 0$ by (2). But this is impossible, since $1 = \|v_n\|_T = \|v_n\| + \|Tv_n\| \to 0$.

Proof of (ii). $T + B$ is closed by (*i*) and Lemma V.3.5.

Proof of (iii). Any set which is $T + B$–bounded is T-bounded, since (*i*) implies the existence of a constant K such that for all

$$f \in \mathfrak{D}(T) = \mathfrak{D}(T + B)$$

$$\|Tf\| \leq \|(T + B)f\| + \|Bf\| \leq \|(T + B)f\| + K\|f\| + \tfrac{1}{2}\|Tf\|$$

whence

$$\|f\|_T \leq 2(K + 1)\|f\|_{T+B}$$

Therefore the T-compactness of B implies the $T + B$–compactness of B. Thus if $T + B$ is normally solvable with $\alpha(T + B) < \infty$, then $T = (T + B) - B$ is also normally solvable and $\kappa(T + B) = \kappa(T)$ by Theorem V.3.7. In particular, $\alpha(T) < \infty$. The converse statement is a particular case of Theorem V.3.7.

The next corollary generalizes a result due to Beals [2].

V.3.9 Corollary. *Let X and Y be complete and let T be a Fredholm operator with domain dense in X. If B is T-strictly singular and B' is T'-strictly singular, then $(T + B)' = T' + B'$.*

Proof. We first note that if U and V are linear operators with equal, finite indices and V is an extension of U, then $U = V$. This follows from the relations

$$\alpha(U) \leq \alpha(V) \qquad \beta(U) \geq \beta(V) \qquad \text{and} \qquad \kappa(U) = \kappa(V)$$

which imply $\alpha(U) = \alpha(V)$ and $\beta(U) = \beta(V)$.

Now, $(T + B)'$ is clearly an extension of $T' + B'$. Moreover, it follows from Theorems V.3.7 and IV.2.3 that

$$\kappa((T + B)') = -\kappa(T + B) = -\kappa(T) = \kappa(T') = \kappa(T' + B')$$

Thus $(T + B)' = T' + B'$.

chapter VI

APPLICATIONS TO ORDINARY DIFFERENTIAL OPERATORS

In this chapter we shall consider properties of certain linear operators which arise from ordinary differential expressions of the form

$$\tau = a_n D^n + a_{n-1} D^{n-1} + \cdots + a_1 D + a_0$$

where $D = d/dt$ and the coefficients a_k are complex-valued functions of a real variable, subject to certain conditions on an interval I.

A differential expression τ may give rise to many operators T which have their domains in $\mathfrak{L}_p(I)$ and ranges in $\mathfrak{L}_q(I)$, where $1 \leq p, q \leq \infty$. For example, T may be defined as follows:

$$\mathfrak{D}(T) = \{f \mid f \in C^\infty(I) \cap \mathfrak{L}_p(I), \tau f \in \mathfrak{L}_q(I)\}$$

$$Tf = \tau f = \sum_{k=0}^{n} a_k D^k f$$

Thus the operators determined by τ are distinguished by means of their domains. The ones which we present are those which are of recognized importance and to which the general theory developed in the preceding chapters applies.

When there is no possibility of confusion, a function in $\mathfrak{L}_p{}^0(I)$ and the element in $\mathfrak{L}_p(I)$ which it determines are used interchangeably. *Unless mention is made regarding the nature of I, the interval is arbitrary.*

The main results, with somewhat different proofs, in Secs. VI.1 to VI.5, are due to Rota [1] for $1 \leq p = q \leq \infty$.

VI.1 CONJUGATES AND PRECONJUGATES OF DIFFERENTIAL OPERATORS

In order to determine some properties of a differential operator T with domain a subspace of, say, $\mathfrak{L}_\infty(I)$ and range in $\mathfrak{L}_q(I)$, $1 < q \leq \infty$, we introduce what we shall call the preconjugate of T. Since the domains of the operators T which we consider are not dense in $\mathfrak{L}_\infty(I)$, the preconjugate $'T$ of T is introduced to compensate for the nonexistence of T'. The idea is to consider T as a map from a subspace of the conjugate space $\mathfrak{L}_\infty(I) = \mathfrak{L}_1'(I)$ into the conjugate space $\mathfrak{L}_q(I) = \mathfrak{L}_{q'}'(I)$ and to define $'T$ in such a way that, in certain cases, T is the conjugate of $'T$.

VI.1.1 Definition. *Let X and Y be normed linear spaces and let T be a linear operator with domain a total subspace of Y' and range in X'. The* ***preconjugate*** *$'T$ of T is defined as follows. $\mathfrak{D}('T)$ is the set of those $x \in X$ for which there exists a $y \in Y$ such that $Ty'x = y'y$ for all y' in $\mathfrak{D}(T)$. Let $'Tx = y$.*

Since $\mathfrak{D}(T)$ is total, $'T$ is unambiguously defined. It follows that $'T$ is a linear map from a subspace of X into Y and

$$Ty'x = y'('Tx) \qquad x \in \mathfrak{D}('T),\ y' \in \mathfrak{D}(T)$$

VI.1.2 Lemma. *Let T be a linear operator with domain a total subspace of Y' and range in X'. Then the preconjugate of T is closed.*

Proof. Suppose $x_n \to x$ and $'Tx_n \to y$. Then for each $y' \in \mathfrak{D}(T)$, $Ty'x_n \to Ty'x$ and $Ty'x_n = y'('Tx_n) \to y'y$. Thus $Ty'x = y'y$, $y' \in \mathfrak{D}(T)$, which means that $x \in \mathfrak{D}('T)$ and $'Tx = y$.

VI.1.3 Lemma. *Let T be a linear operator with domain a total subspace of Y' and range in X'. Suppose $\mathfrak{D}('T)$ is dense in X. Then $('T)'$ is an extension of T, and therefore T is closable. If, in addition, X and Y are reflexive, then $\bar{T} = ('T)'$, where $\bar{T}$ is the minimal closed linear extension of T.*

Proof. If $y' \in \mathfrak{D}(T)$ and $x \in \mathfrak{D}('T)$, then $y'('Tx) = Ty'x$. Thus $y' \in \mathfrak{D}(('T)')$ and $('T)'y' = Ty'$, $y' \in \mathfrak{D}(T)$. Since the conjugate operator $('T)'$ is closed, T is closable.

Let X and Y be reflexive. Suppose $\bar{T} \neq ('T)'$. Then there exists some $(v', ('T)'v') \notin G(\bar{T}) = \overline{G(T)}$. Thus there exists a $z'' \in (Y' \times X')'$

such that $z''(v', ('T)'v') \neq 0$, and $z''(y', Ty') = 0$, $y' \in \mathfrak{D}(T)$. Define $y'' \in Y''$ and $x'' \in X''$ by $y''y' = z''(y', 0)$ and $x''x' = z''(0, x')$. Then

$$y''y' + x''Ty' = z''(y', Ty') = 0 \qquad y' \in \mathfrak{D}(T) \tag{1}$$

$$y''v' + x''('T)'v' \neq 0 \tag{2}$$

Thus, since X and Y are reflexive, there exist $x \in X$ and $y \in Y$ such that

$$y'y + Ty'x = 0 \qquad y' \in \mathfrak{D}(T) \tag{3}$$

$$v'y + ('T)'v'x \neq 0 \tag{4}$$

Now, (3) shows that x is in $\mathfrak{D}('T)$ and $'Tx = -y$. But then (4) implies that $v'y + v'('Tx) = v'y - v'y \neq 0$, which is absurd. Hence $\bar{T} = ('T)'$.

VI.1.4 Lemma. *Let A be a closable linear operator with domain dense in X and range in Y. Then $\bar{A} = '(A')$.*

Proof. Since A is closable, $\mathfrak{D}(A')$ is total by Theorem II.2.11. Hence the preconjugate of A' is defined. Now, $'(A')$ is an extension of A; for if $x \in \mathfrak{D}(A)$, then $A'y'x = y'Ax$, $y' \in \mathfrak{D}(A')$. Thus, by the definition of $'(A')$, x is in $\mathfrak{D}('(A'))$ and $'(A')x = Ax$. Since $'(A')$ is closed by Lemma VI.1.2, it remains to prove that $G('(A')) \subset G(\bar{A}) = \overline{G(A)}$. By an argument analogous to the one given in Lemma VI.1.3, it follows that $G('(A')) \subset G(\bar{A})$.

VI.1.5 Lemma. *If T is the conjugate of a closable operator, then $T = ('T)'$.*

Proof. Suppose $T = A'$, where A is closable. Then, by Lemma VI.1.4, $\bar{A} = '(A') = 'T$. Hence

$$T = A' = (\bar{A})' = ('T)'$$

by Theorem II.2.11.

VI.1.6 Definition. *Let τ be a differential expression of the form*

$$\tau = \sum_{k=0}^{n} a_k D^k$$

where each a_k is a complex-valued function on I. For each positive integer n, define $A_n(I)$ to be the set of complex-valued functions f on I for which $f^{(n-1)} = D^{n-1}f$ exists and is absolutely continuous on every compact subinterval of I Let $A_0(I) = C(I)$.

*The **maximal** operator $T_{\tau,p,q}$, $1 \leq p,\ q \leq \infty$, corresponding to τ, p, q and I, is defined as follows.*

$$\mathfrak{D}(T_{\tau,p,q}) = \{f \mid f \in A_n(I) \cap \mathfrak{L}_p(I),\ \tau f \in \mathfrak{L}_q(I)\}$$

$$T_{\tau,p,q}f = \tau f = \sum_{k=0}^{n} a_k D^k f$$

The operator $T^R_{\tau,p,q}$ is defined to be the restriction of $T_{\tau,p,q}$ to those $f \in \mathfrak{D}(T_{\tau,p,q})$ which have compact support in the interior of I.

Since an absolutely continuous function is differentiable a.e., it follows that τf is defined a.e. on I for $f \in A_n(I)$.

Before proving Theorem VI.1.9, we make the following observations which indicate what one might expect $(T^R_{\tau,p,q})'$ and $'(T^R_{\tau,p,q})$ to be.

Suppose $\tau = \sum_{k=0}^{n} a_k D^k$, $a_k \in C^k(I)$, $0 \leq k \leq n$. Since $C_0^\infty(I)$ is contained in $\mathfrak{D}(T^R_{\tau,p,q})$, we know, by Theorem 0.9, that $\mathfrak{D}(T^R_{\tau,p,q})$ is dense in $\mathfrak{L}_p(I)$ when $1 \leq p < \infty$ and is total in $\mathfrak{L}^\infty(I)$. Thus $T^R_{\tau,p,q}$ has a conjugate $(T^R_{\tau,p,q})'$, defined with domain a subspace of $\mathfrak{L}'_q(I) = \mathfrak{L}_{q'}(I)$ and range in $\mathfrak{L}'_p(I) = \mathfrak{L}_{p'}(I)$, where $1 \leq p,\ q < \infty$, p' and q' conjugate to p and q, respectively. For $1 < p,\ q \leq \infty$, $T^R_{\tau,p,q}$ may be considered as a map with domain a total subspace of the conjugate space $\mathfrak{L}_p(I) = \mathfrak{L}'_{p'}(I)$ and range in the conjugate space $\mathfrak{L}_q(I) = \mathfrak{L}'_{q'}(I)$. Hence the preconjugate $'T^R_{\tau,p,q}$ of $T^R_{\tau,p,q}$ is defined as a map from a subspace of $\mathfrak{L}_{q'}(I)$ to $\mathfrak{L}_{p'}(I)$.

To simplify the discussion, we assume that $I = [a, b]$ is compact. Writing $T_R = T^R_{\tau,p,q}$, let g be in $\mathfrak{D}(T'_R)$ ($g \in \mathfrak{D}('T_R)$) and let

$$T'_R g = g^* \qquad ('T_R g = g^*)$$

Then for $f \in \mathfrak{D}(T_R)$,

$$g^*(f) = (T'_R g)f = g(T_R f) \qquad ((T_R f)g = f('T_R g) = f(g^*))$$

Thus, regardless of whether g is in $\mathfrak{D}(T'_R)$ or in $\mathfrak{D}('T_R)$,

$$\text{(1)} \qquad \int_a^b g^* f = \int_a^b g\tau f = \int_a^b \sum_{k=0}^{n} (a_k g) f^{(k)} \qquad f \in \mathfrak{D}(T_R)$$

In order to obtain an expression for g^*, equation (1) suggests successive integration by parts of each $\int_a^b (a_k g) f^{(k)}$ in order to shift the kth derivative of f to that of $a_k g$. Thus we further assume that g is in $A_n(I)$.

Since a_k is in $C^k(I)$, $1 \leq k \leq n$, $(a_k g)^{(k-1)}$ is absolutely continuous. Hence integration by parts gives for $1 \leq k \leq n$,

$$(2) \qquad \begin{aligned} \int_a^b (a_k g) f^{(k)} &= (a_k g) f^{(k-1)}\Big]_a^b - \int_a^b (a_k g)' f^{(k-1)} \\ &= (a_k g) f^{(k-1)} - (a_k g)' f^{(k-2)}\Big]_a^b + \int_a^b (a_k g)^{(2)} f^{(k-2)} \\ &= \sum_{j=0}^{k-1} (-1)^j (a_k g)^{(j)} f^{(k-1-j)}\Big]_a^b + \int_a^b (-1)^k (a_k g)^{(k)} f \end{aligned}$$

where $v\Big]_a^b$ denotes $v(b) - v(a)$ for any function v defined at a and b. From (1) and (2),

$$(3) \qquad \int_a^b g\tau f = \sum_{k=1}^{n} \sum_{j=0}^{k-1} (-1)^j (a_k g)^{(j)} f^{(k-1-j)}\Big]_a^b + \int_a^b \sum_{k=0}^{n} (-1)^k (a_k g)^{(k)} f$$

Since $f \in \mathfrak{D}(T_R)$ has compact support in the interior of I,

$$f^{(k)}(a) = f^{(k)}(b) = 0 \qquad 0 \leq k \leq n-1$$

Hence, from (1) and (3),

$$(4) \qquad \int_a^b g^* f = \int_a^b (\tau^* g) f$$

where $\tau^* g$ is the function defined a.e. by

$$\tau^* g = \sum_{k=0}^{n} (-1)^k (a_k g)^{(k)}$$

Since $\mathfrak{D}(T_R)$ is dense in $\mathfrak{L}_p(I)$, $1 \leq p < \infty$, and total in $\mathfrak{L}_1'(I) = \mathfrak{L}_\infty(I)$, equation (4) implies $g^* = \tau^* g$ a.e. on I. Hence

$$T_R' g = \tau^* g \qquad ('T_R g = \tau^* g) \qquad g \in A_n(I)$$

Thus τ^* appears to determine the conjugate and preconjugate of T_R in a manner similar to that in which τ determines T_R.

VI.1.7 Definition. *Let* $\tau = \sum_{k=0}^{n} a_k D^k$, $a_k \in C^k(I)$, $0 \leq k \leq n$. *The Lagrange, or formal, adjoint* τ^* *of* τ *is the expression defined by*

$$\tau^* g = \sum_{k=0}^{n} (-1)^k D^k (a_k g)$$

If $g^{(n)}$ exists a.e., then $(\tau^* g)(t) = \sum_{k=0}^{n} (-1)^k (a_k g)^{(k)}(t)$ exists for almost all t.

The following lemma was proved in the above discussion.

VI.1.8 Lemma. *Let* $\tau = \sum_{k=0}^{n} a_k D^k$, $a_k \in C^k(I)$, $0 \leq k \leq n$. *If* f *and* g *are in* $A_n(I)$, *where* $I = [a, b]$ *is compact, then*

$$\int_a^b g\tau f = \sum_{k=1}^{n} \sum_{j=0}^{k-1} (-1)^j (a_k g)^{(j)} f^{(k-1-j)} \Big]_a^b + \int_a^b f\tau^* g$$

The next theorem was proved by Halperin [1] for $p = q = 2$ and by Rota [1] for $1 \leq p = q \leq \infty$.

VI.1.9 Theorem. *Let* τ *be the differential expression* $\tau = \sum_{k=0}^{n} a_k D^k$, *where* $a_k \in C^k(I)$, $0 \leq k \leq n$, *and* $a_n(t) \neq 0$, $t \in I$. *Then*

$$(T^R_{\tau,p,q})' = T_{\tau^*,q',p'} \qquad 1 \leq p, q < \infty$$

$$'T^R_{\tau,p,q} = T_{\tau^*,q',p'} \qquad 1 < p, q \leq \infty$$

Proof. Write $T_R = T^R_{\tau,p,q}$ and $T_* = T_{\tau^*,q',p'}$. Suppose $g \in \mathfrak{D}(T_*)$ and $f \in \mathfrak{D}(T_R)$. Then f vanishes outside of some compact interval $[a, b]$ contained in the interior of I. Therefore we have, by Lemma VI.1.8,

$$\int_I g\tau f = \int_a^b g\tau f = \int_a^b (\tau^* g) f = \int_I (\tau^* g) f \tag{1}$$

If $1 \leq p, q < \infty$, then (1) shows that

$$(T_* g) f = g T_R f \qquad f \in \mathfrak{D}(T_R)$$

Hence $g \in \mathfrak{D}(T'_R)$ and $T'_R g = T_* g$.

If $1 < p, q \leq \infty$, then considering T_R as an operator from $\mathfrak{L}'_{p'} = \mathfrak{L}_p(I)$ into $\mathfrak{L}'_{q'} = \mathfrak{L}_q(I)$ and f as a functional in $\mathfrak{L}'_{p'}$, it follows from (1) that

$$f(T_* g) = (T_R f) g \qquad f \in \mathfrak{D}(T_R)$$

Hence $g \in \mathfrak{D}('T_R)$ and $'T_R g = T_* g$. To prove the theorem, it remains to show $\mathfrak{D}(T'_R) \subset \mathfrak{D}(T_*)$, $1 \leq p$, $q < \infty$, and $\mathfrak{D}('T_R) \subset \mathfrak{D}(T_*)$, $1 < p$, $q \leq \infty$.

Suppose $g \in \mathfrak{D}(T'_R)$ $(g \in \mathfrak{D}('T_R))$ and $T'_R g = g^*$ $('T_R g = g^*)$. Then for every $f \in \mathfrak{D}(T_R)$,

$$\int_I (\tau f) g = \int_I f g^* \tag{2}$$

To show that g is in $\mathfrak{D}(T_*)$, it suffices to show that for every compact interval $I_0 = [a, b] \subset I$, g, considered as a function, is equal a.e. to a function in $A_n(I_0)$ and $\tau^* g = g^*$ a.e. on I_0. The remark follows from the observations that any interval is the union of an ascending sequence of compact subintervals and that continuous functions which are equal a.e. are identical.

Let $D_0 = \{f \mid f \in \mathfrak{D}(T_R),\ f \text{ has support in } I_0\}$. Since elements in $\mathfrak{D}(T_R)$ have support in the interior of I, it follows that, given $f \in D_0$, $f^{(k)}(a) = f^{(k)}(b) = 0$, $0 \le k \le n - 1$. Successive integration by parts yields the formula

$$f^{(k)}(t) = \int_a^t \frac{(t - s)^{n-k-1}}{(n - k - 1)!} f^{(n)}(s)\, ds \qquad 0 \le k \le n - 1 \tag{3}$$

Now, f vanishes outside of I_0. Hence (2) and (3) imply

$$\begin{aligned} \int_a^b a_n(s) g(s) f^{(n)}(s)\, ds + \sum_{k=0}^{n-1} \int_a^b dt \int_a^t a_k(t) g(t) \frac{(t - s)^{n-k-1}}{(n - k - 1)!} f^{(n)}(s)\, ds \\ = \int_a^b dt \int_a^t g^*(t) \frac{(t - s)^{n-1}}{(n - 1)!} f^{(n)}(s)\, ds \end{aligned} \tag{4}$$

All the integrands in (4) which involve both s and t are in $\mathfrak{L}_1(I_0 \times I_0)$, since a_k is continuous on I_0, $g \in \mathfrak{L}_{q'}(I_0) \subset \mathfrak{L}_1(I_0)$, and $f^{(n)} \in \mathfrak{L}_1(I_0)$. Thus, by Fubini's theorem, we may change the order of integration in (4) and obtain

$$\begin{aligned} 0 = \int_a^b f^{(n)}(s) \Bigg[a_n(s) g(s) + \sum_{k=0}^{n-1} \int_s^b \frac{(t - s)^{n-k-1}}{(n - k - 1)!} a_k(t) g(t)\, dt \\ - \int_s^b \frac{(t - s)^{n-1}}{(n - 1)!} g^*(t)\, dt \Bigg] ds \end{aligned} \tag{5}$$

for all $f \in D_0$.

Let $F(s)$ be the expression inside the square brackets in (5). We now show that F is equivalent on I_0 to a polynomial of degree at most $n - 1$.

Given $Q \in \mathfrak{L}_q(I_0)$ such that Q is orthogonal to the subspace $\mathcal{P}$ of $\mathfrak{L}_{q'}(I_0)$ consisting of the polynomials of degree at most $n - 1$, the function

h defined by

$$h(t) = \int_a^t \frac{(t-s)^{n-1}}{(n-1)!} Q(s)\,ds \qquad t \in I_0$$

$$h(t) = 0 \qquad t \notin I_0$$

is easily seen to be in D_0 with $h^{(n)}(t) = Q(t)$ a.e. on I_0. Thus (5) holds for $h^{(n)}$ in place of $f^{(n)}$, whence

$$0 = \int_a^b Q(s)F(s)\,ds \tag{6}$$

for all $Q \in \mathfrak{L}_q(I_0)$ orthogonal to $\mathcal{P} \subset \mathfrak{L}_{q'}(I_0)$; that is, (6) holds for all $Q \in \mathcal{P}^\perp$ when $1 < q \leq \infty$ or for all $Q \in {}^\perp\mathcal{P}$ when $q = 1$. Hence, Theorem II.3.4, Remark II.3.6, and the finite dimensionality of $\mathcal{P}$ imply

$$F \in {}^\perp(\mathcal{P}^\perp) = \mathcal{P} \qquad 1 < q \leq \infty$$

$$F \in ({}^\perp\mathcal{P})^\perp = \mathcal{P} \qquad q = 1$$

Thus F is equivalent on I_0 to a polynomial v of degree at most $n - 1$. Therefore, except for a set of Lebesgue measure zero,

$$a_n(s)g(s) = v(s) - \sum_{k=0}^{n-1} \int_s^b \frac{(t-s)^{n-k-1}}{(n-k-1)!} a_k(t)g(t)\,dt + \int_s^b \frac{(t-s)^{n-1}}{(n-1)!} g^*(t)\,dt \tag{7}$$

Now the right hand side of (7) and $1/a_n$ are absolutely continuous on I_0. Thus we can redefine g on a set of measure zero to be absolutely continuous on I_0 so that (7) holds on all of I_0. Differentiating both sides of (7), we obtain for almost all $s \in I_0$

$$\begin{aligned}(a_n g)'(s) &= a_n(s)g'(s) + g(s)a_n'(s) \\ &= v'(s) + a_{n-1}(s)g(s) + \sum_{k=0}^{n-2} \int_s^b \frac{(t-s)^{n-k-2}}{(n-k-2)!} a_k(t)g(t)\,dt \\ &\qquad - \int_s^b \frac{(t-s)^{n-2}}{(n-2)!} g^*(t)\,dt\end{aligned} \tag{8}$$

Since g, a_n' and $1/a_n$ are absolutely continuous, it follows from (8) that g' is equivalent to an absolutely continuous function. But then g' is absolutely continuous on I_0; for if the derivative g' of an absolutely continuous function g is equivalent to a continuous function h, then g' is identically h. This follows from

$$g(x) - g(a) = \int_a^x g'(t)\,dt = \int_a^x h(t)\,dt$$

Repeated differentiation of both sides of the equalities in (8), together with a repetition of the argument just used to prove g' is absolutely continuous, shows that $g^{(n-1)}$ is absolutely continuous and $\tau^*g = g^*$ a.e. on I_0. (Recall that v is a polynomial of degree at most $n - 1$.) Thus the proof of the theorem is concluded.

Before proving the next corollary, we take a closer look at τ^*.

VI.1.10 Lemma. Let $\tau = \sum_{k=0}^{n} a_k D^k$, $a_k \in C^k(I)$, $0 \le k \le n$. *Then for* $g \in A_n(I)$

$$\tau^*g = \sum_{k=0}^{n} b_k g^{(k)} \text{ a.e.}$$

where

$$b_k = \sum_{j=k}^{n} (-1)^j \binom{j}{k} D^{j-k} a_j \qquad 0 \le k \le n$$

For $0 \le k \le n$, b_k *is in* $C^k(I)$.

Proof. Let t be such that $g^{(n)}(t)$ exists. Then

$$(\tau^*g)(t) = \sum_{k=0}^{n} (-1)^k (a_k g)^{(k)}(t) \tag{1}$$

By Leibnitz's rule,

$$(a_j g)^{(j)}(t) = \sum_{i=0}^{j} \binom{j}{i} a_j^{(j-i)}(t) g^{(i)}(t) \qquad 1 \le j \le n \tag{2}$$

Setting $i = k$ and $j = k, k + 1, \ldots, n$ in (2) implies, together with (1), that

$$(\tau^*g)(t) = \sum_{k=0}^{n} b_k(t) g^{(k)}(t) \qquad b_k(t) = \sum_{j=k}^{n} (-1)^j \binom{j}{k} a_j^{(j-k)}(t)$$

Since $D^k a_j^{(j-k)} = a_j^{(j)}$ and a_j is in $C^j(I)$, $0 \le k \le j \le n$, it follows that b_k is in $C^k(I)$, $0 \le k \le n$.

VI.1.11 Remark. If τ^* is identified with $\sum_{k=0}^{n} b_k D^k$ as given in the preceding lemma, then the Lagrange adjoint τ^{**} of τ^* is defined. In the same way, τ^{**} can be identified with $\sum_{k=0}^{n} c_k D^k$, where

$$c_k = \sum_{j=k}^{n} (-1)^j \binom{j}{k} b_j^{(j-k)}$$

VI.1.12 Lemma. *Let c_k be as in Remark* VI.1.11. *Then c_k is identically a_k on I for $0 \leq k \leq n$. Thus $\tau^{**} = \tau$.*

Proof. Let $I_1 = [a, b]$ be any compact subinterval of the interior of I. Suppose g is in $C^\infty(I_1)$ and f is in $C_0^\infty(I_1)$. Upon replacing τ by $\tau^* = \sum_{k=0}^{n} b_k D^k$ in Lemma VI.1.8 and noting that $0 = f^{(k)}(a) = f^{(k)}(b)$, $0 \leq k \leq n - 1$, it follows that

$$\int_a^b f\tau g = \int_a^b (\tau^* f)g = \int_a^b f\tau^{**}g$$

Since $C_0^\infty(I_1)$ is dense in $\mathfrak{L}_1(I_1)$ and both τg and $\tau^{**}g$ are continuous on I_1, it follows that $\tau g = \tau^{**}g$. By taking g to be the functions

$$g_j(t) = t^j \qquad 0 \leq j \leq n$$

we obtain $a_k = c_k$ on I_1, $0 \leq k \leq n$. Since I_1 was an arbitrary compact subinterval of the interior of I, the lemma follows.

The lemma could also have been proved by computing the derivatives of c_j in terms of the a_i.

VI.1.13 Corollary

$$T_{\tau,p,q} = {}'T^R_{\tau^*,q',p'} \qquad 1 \leq p, q < \infty$$

$$T_{\tau,p,q} = (T^R_{\tau^*,q',p'})' \qquad 1 < p, q \leq \infty$$

$T_{\tau,p,q}$ *is closed when* $1 \leq p, q < \infty$ *or* $1 < p, q \leq \infty$

Proof. Since $\tau^{**} = \tau$, $p'' = p$, and $q'' = q$, the corollary follows from Theorem VI.1.9 and the fact that conjugate and preconjugate operators are closed.

VI.1.14 Corollary. *The conjugate of $T_{\tau,p,q}$ is the minimal closed extension of $T^R_{\tau^*,q',p'}$ when $1 < p, q < \infty$.*

Proof. By Corollary VI.1.13 and Lemma VI.1.3,

$$T'_{\tau,p,q} = ('T^R_{\tau^*,q',p'})' = \overline{T^R_{\tau^*,q',p'}}$$

VI.2 MINIMAL AND MAXIMAL OPERATORS

Throughout this section, $\tau = \sum_{k=0}^{n} a_k D^k$, where

$$a_k \in C^k(\mathrm{I}) \qquad 0 \le k \le n$$

$$a_n(t) \ne 0,\ t \in \mathrm{I}$$

VI.2.1 Definition. *The minimal operator corresponding to* (τ, p, q) *denoted by* $T_{0,\tau,p,q}$, *is defined to be the minimal closed extension of* $T^R_{\tau,p,q}$ *when* $1 \le p, q < \infty$. *When* $1 < p, q \le \infty$, *the minimal operator is defined to be* $T'_{\tau^*,q',p'}$.

VI.2.2 Remarks

i. There is no ambiguity in the definition of the minimal operator since it follows from Corollary VI.1.14 that

$$T'_{\tau^*,q',p'} = \overline{T^R_{\tau,p,q}} \qquad 1 < p, q < \infty$$

The reason for the seemingly unnatural definition of the minimal operator for the case when p and q are not both finite will be made clear in subsequent theorems.

ii. The maximal operator is an extension of the minimal operator; for when $1 \le p, q < \infty$, then $T_{0,\tau,p,q} = \overline{T^R_{\tau,p,q}} \subset T_{\tau,p,q}$, since $T_{\tau,p,q}$ is closed. When $1 < p, q \le \infty$, it follows from Theorem VI.1.9 that

$$(1) \qquad T_{0,\tau,p,q} = T'_{\tau^*,q',p'} = ('T^R_{\tau,p,q})' \subset ('T_{\tau,p,q})'$$

Now $T_{\tau,p,q} = (T^R_{\tau^*,q',p'})'$ by Corollary VI.1.13. Hence we can conclude from Lemma VI.1.5 that $('T_{\tau,p,q})' = T_{\tau,p,q}$. Thus $T_{0,\tau,p,q} \subset T_{\tau,p,q}$ by (1).

From the definition of the minimal operator, Theorem VI.1.9, and Corollary VI.1.13, the following relationships are obtained.

VI.2.3 Theorem. *For* $1 \le p, q < \infty$,

i. $T'_{0,\tau,p,q} = (\overline{T^R_{\tau,p,q}})' = (T^R_{\tau,p,q})' = T_{\tau^*,q',p'}$

ii. $T_{0,\tau^*,q',p'} = T'_{\tau,p,q}$

For $1 < p, q \le \infty$,

$$iii. \quad T_{0,\tau,p,q} = T'_{\tau^*,q',p'}$$
$$iv. \quad T'_{0,\tau^*,q',p'} = (T^R_{\tau^*,q',p'})' = T_{\tau,p,q}$$

VI.2.4 Lemma. *Suppose y is in $A_n(I)$, $\tau y = 0$ a.e., and*

$$0 = y(a) = y'(a) = \cdot \cdot \cdot = y^{(n-1)}(a)$$

or some $a \in I$. Then $y = 0$ on I.

Proof. Suppose $y(x_1) \neq 0$ for some $x_1 \in I$. For definiteness, assume $x_1 > a$. On $[a, x_1]$ define

$$s(x) = \sum_{k=0}^{n-1} |y^{(k)}(x)|$$

and

$$x_0 = \sup \{x \mid s(x) = 0 \qquad x \in [a, x_1]\}$$

Since s is continuous, $s(x_0) = 0$, or equivalently,

$$y^{(k)}(x_0) = 0 \qquad 0 \le k \le n - 1$$

Thus

$$(1) \quad y^{(k)}(x) = y^{(k)}(x) - y^{(k)}(x_0) = \int_{x_0}^{x} y^{(k+1)}(t)\, dt \qquad 0 \le k \le n - 1$$

and

$$(2) \qquad y^{(n-1)}(x) = \int_{x_0}^{x} y^{(n)}(t)\, dt = - \int_{x_0}^{x} \sum_{k=0}^{n-1} \frac{a_k(t) y^{(k)}(t)}{a_n(t)}\, dt$$

Letting $M(x) = \sup\limits_{x_0 \le t \le x} s(t)$, (1) and (2) imply

$$(3) \qquad 0 < M(x) \le M(x) \int_{x_0}^{x} \sum_{k=0}^{n-1} \left(\left| \frac{a_k(t)}{a_n(t)} \right| + 1 \right) dt \qquad x > x_0$$

But this cannot be, since the integral in (3) is less than 1 when x is sufficiently close to x_0. Thus $y = 0$ on I. The proof for $x_1 < a$ is similar.

VI.2.5 Theorem. $\dim \mathfrak{N}(T_{\tau,p,q}) \le n$, $1 \le p, q \le \infty$.

Proof. Suppose $\{y_1, y_2, \ldots, y_n\}$ is a linearly independent subset of $\mathfrak{N}(T_{\tau,p,q})$. For $a \in I$, the determinant of the matrix $(y_j^{(k)}(a))$, $0 \le$

$k \leq n - 1$, is not zero; otherwise there exists a set of numbers $c_1, c_2, \ldots, c_n$, not all zero, such that

$$0 = \sum_{i=1}^{n} c_i y_i^{(k)}(a) \qquad k = 0, 1, \ldots, n-1$$

Hence $y = \sum_{i=1}^{n} c_i y_i$ satisfies the hypotheses of Lemma VI.2.4 and is therefore 0 on I. This contradicts the assumption that $\{y_1, y_2, \ldots, y_n\}$ is linearly independent. Thus, given $g \in \mathfrak{N}(T_{\tau,p,q})$, we know from Cramer's rule that the system of equations

$$g^{(k)}(a) = \sum_{i=1}^{n} y_i^{(k)}(a) x_i \qquad k = 0, 1, \ldots, n-1$$

has a solution $x_i = b_i$, $i = 1, 2, \ldots, n$. Therefore $g - \sum_{i=1}^{n} b_i y_i$ satisfies the hypotheses of Lemma VI.2.4. Hence $g = \sum_{i=1}^{n} b_i y_i$, which proves that $\dim \mathfrak{N}(T_{\tau,p,q})$ cannot exceed n.

Note that the proofs of Lemma VI.2.4 and Theorem VI.2.5 also hold even when a_k is locally integrable, $0 \leq k \leq n - 1$, and $1/a_n$ is locally in $\mathfrak{L}_\infty(I)$.

VI.2.6 Definition. *A pair of numbers (p, q) is called* ***admissible*** *if $1 \leq p, q < \infty$ or $1 < p, q \leq \infty$.*

VI.2.7 Theorem. *Let (p, q) be admissible. If any one of the minimal or maximal operators corresponding to (τ, p, q) or to (τ^*, q', p') has a closed range, then all four of the operators are Fredholm operators and*

$$\dim \frac{\mathfrak{D}(T_{\tau,p,q})}{\mathfrak{D}(T_{o,\tau,p,q})} = \kappa(T_{\tau,p,q}) - \kappa(T_{o,\tau,p,q})$$

Proof. For convenience we write

$$T = T_{\tau,p,q}$$

$$T_o = T_{o,\tau,p,q}$$

$$T_* = T_{\tau^*,q',p'}$$

$$T_{*o} = T_{o,\tau^*,q',p'}$$

If either T or T_* has a closed range, it follows from Theorems VI.2.3 and IV.1.2 that either T_{*o} or T_o has a closed range. Thus we need only consider the cases when T_o or T_{*o} have closed range. Suppose T_o has a closed range.

If $1 \le p, q < \infty$, then by Theorems IV.2.3, VI.2.3, and VI.2.5,

$$\beta(T) \le \beta(T_o) = \alpha(T_o') = \alpha(T_*) \le n$$

Thus T and T_o are Fredholm operators. Similarly, by Theorem VI.2.3,

$$\beta(T_*) \le \beta(T_{*o}) = \beta(T') = \alpha(T) \le n$$

Hence we have shown that T_o, T, T_{*o}, and T_* are Fredholm operators when T_o has a closed range and $1 \le p, q < \infty$.

If $1 < p, q \le \infty$, then from Theorem VI.2.3 it follows that

$$\beta(T) \le \beta(T_o) = \beta(T_*') = \alpha(T_*) \le n$$

$$\beta(T_*) \le \alpha(T_*') = \alpha(T_o) \le \alpha(T) \le n$$

Thus T_o, T, T_{*o}, and T_* are Fredholm operators when $\mathcal{R}(T_o)$ is closed and (p, q) is admissible.

If $\mathcal{R}(T_{*o})$ is closed, then the above results applied to T_{*o} in place of T_o and T_* in place of T prove that T_{*o}, T_*, T_o, and T are Fredholm operators when (p, q) is admissible.

Finally, we note that

$$\text{(1)} \qquad \dim \frac{\mathfrak{D}(T)}{\mathfrak{D}(T_o) + \mathfrak{N}(T)} \le \dim \frac{Y}{\mathcal{R}(T_o)} = \beta(T_o) < \infty$$

since the linear mapping $\nu: \mathfrak{D}(T)/(\mathfrak{D}(T_o) + \mathfrak{N}(T)) \to Y/\mathcal{R}(T_o)$, defined by $\nu[x] = [Tx]$, is 1-1. Since $\alpha(T) < \infty$, (1) implies $\dim \mathfrak{D}(T)/\mathfrak{D}(T_o) < \infty$ and therefore, by Lemma V.1.5,

$$\kappa(T) - \kappa(T_o) = \dim \frac{\mathfrak{D}(T)}{\mathfrak{D}(T_o)}$$

VI.2.8 Lemma. *Given compact interval $[a, b]$ and numbers $c_0, c_1, \ldots,$ c_{n-1}, there exists a function g which is infinitely differentiable on the line such that*

$$g^{(k)}(a) = c_k \qquad k = 0, 1, \ldots, n-1$$

$$g(x) = 0 \qquad x \ge b$$

Proof. Let P be any polynomial such that $P^{(k)}(a) = c_k$, $k = 0, 1, \ldots, n-1$; for example, $P(x) = \sum_{k=0}^{n-1} \frac{c_k}{k!}(x-a)^k$. By Theorem 0.8 there exists a φ which is infinitely differentiable on the line, has support in $[a-1, b]$, and is identically 1 on an interval containing a as an interior point. Define $g(x) = p(x)\varphi(x)$. Then by Leibnitz's rule,

$$g^{(k)}(x) = \sum_{i=0}^{k} \binom{k}{i} p^{(k-i)}(x)\varphi^{(i)}(x) \qquad k = 0, 1, \ldots, n-1$$

from which the lemma follows.

VI.2.9 Lemma. *If I contains one of its endpoints a, then every f in the domain of the minimal operator corresponding to (τ, p, q), (p, q) admissible, satisfies the conditions*

$$0 = f(a) = f'(a) = \ldots = f^{(n-1)}(a)$$

Proof. For convenience we write $T_o = T_{o,\tau,p,q}$, $T_R = T^R_{\tau,p,q}$, $T_* = T_{\tau^*,q',p'}$. Since $T_o' = T_*$ for $1 \leq p, q < \infty$ and $T_o = T_*'$ for $1 < p, q \leq \infty$, it follows that for $f \in \mathfrak{D}(T_o)$,

$$(1) \qquad 0 = \int_I f\tau^*g - \int_I (\tau f)g \qquad g \in \mathfrak{D}(T_*)$$

Let $[a, b]$ be a compact subinterval of I. For any g in $\mathfrak{D}(T_*)$ such that $g(x) = 0$ for all $x \geq b$ in I, we have from (1) and Lemma VI.1.8 that

$$0 = \sum_{k=1}^{n} \sum_{j=0}^{k-1} (-1)^j D^j(a_k g) D^{k-j-1} f \Big]_a^b$$

Since, by Lemma VI.2.8, $g^{(k)}(a)$ may be chosen to have any prescribed values and $g^{(k)}(b) = 0$, it is easy to show that

$$0 = f^{(k)}(a) \qquad k = 0, 1, \ldots, n-1$$

VI.2.10 Theorem. *Let I contain one of its endpoints and let (p, q) be admissible. Then*

i. The minimal operator corresponding to (τ, p, q) is 1-1.
ii. The maximal operator corresponding to (τ, p, q) has range dense in $\mathfrak{L}_q(I)$, $1 \leq p, q < \infty$.

Proof. If $T_{o,\tau,p,q}y = 0$, then $y = 0$ by Lemmas VI.2.9 and VI.2.4. Thus $T_{o,\tau,p,q}$ is 1-1.

Since $T'_{\tau,p,q} = T_{o,\tau^*,q',p'}$, $1 \leq p, q < \infty$, it follows from what was just proved and Theorem II.3.7 that $T_{\tau,p,q}$ has range dense in $\mathcal{L}_q(I)$.

VI.2.11 Theorem. *Suppose I contains one of its endpoints and (p, q) is admissible. If any one of the minimal or maximal operators corresponding to (τ, p, q) or (τ^*, q', p') has a closed range, then $T_{\tau,p,q}$ is surjective and $T_{o,\tau,p,q}$ has a bounded inverse.*

Proof. The theorem follows from Theorems VI.2.10 and VI.2.7 and Lemma IV.1.1.

VI.3 MAXIMAL OPERATORS CORRESPONDING TO COMPACT INTERVALS

VI.3.1 Theorem. *Let $I = [a, b]$ be compact and let T be the maximal operator corresponding to (τ, p, q), $1 \leq p, q \leq \infty$, where τ is the differential expression*

$$\tau = \sum_{k=0}^{n} a_k D^k \qquad a_k \in \mathcal{L}_1(I),\ 0 \leq k \leq n-1$$

$$\frac{1}{a_n} \in \mathcal{L}_\infty(I)$$

Then

i. T is surjective and has kernel index n.

ii. If $f \in \mathcal{L}_q(I)$, $1 \leq q \leq \infty$, then $Ty = f$, where

$$y(t) = \int_a^t G(s, t)\,\frac{f(s)}{a_n(s)}\,ds$$

and G is the continuous function on $I \times I$ expressed as the following quotient of determinants.

$$G(s, t) = \frac{\begin{vmatrix} y_1(s) & y_2(s) & \cdots & y_n(s) \\ y_1'(s) & y_2'(s) & \cdots & y_n'(s) \\ \cdots & \cdots & \cdots & \cdots \\ y_1^{(n-2)}(s) & y_2^{(n-2)}(s) & \cdots & y_n^{(n-2)}(s) \\ y_1(t) & y_2(t) & \cdots & y_n(t) \end{vmatrix}}{\begin{vmatrix} y_1(s) & y_2(s) & \cdots & y_n(s) \\ y_1'(s) & y_2'(s) & \cdots & y_n'(s) \\ \cdots & \cdots & \cdots & \cdots \\ y_1^{(n-1)}(s) & y_2^{(n-1)}(s) & \cdots & y_n^{(n-1)}(s) \end{vmatrix}}$$

with $y_1, y_2, \ldots, y_n$ any basis for $\mathfrak{N}(T)$.

iii. $$\gamma(T) \geq \left\| \frac{1}{a_n} \right\|_\infty^{-1} \left\| \left[\int_a^t |G(s, t)|^{q'}\,ds \right]^{1/q'} \right\|_p^{-1} \qquad \begin{matrix} 1 \leq p \leq \infty \\ 1 < q \leq \infty \end{matrix}$$

$$\gamma(T) \geq \left\| \frac{1}{a_n} \right\|_\infty^{-1} \| \max_{a \leq s \leq t} |G(s, t)| \|_p^{-1} \qquad \begin{matrix} 1 \leq p \leq \infty \\ q = 1 \end{matrix}$$

where $\gamma(T)$ is the minimum modulus of T, $\| \ \|_p$ is the norm on $\mathfrak{L}_p(I)$, and q' is conjugate to q.

Proof of (i). From the condition put on a_n, we may assume, without loss of generality, that $a_n = 1$. We first prove that not only is T surjective, but that given $f \in \mathfrak{L}_1(I)$ and constants $c_0, c_1, \ldots, c_{n-1}$, there exists a $y \in \mathfrak{D}(T)$ such that

$$(1) \qquad \begin{aligned} Ty &= f \\ y^{(k)}(a) &= c_k \qquad 0 \leq k \leq n-1 \end{aligned}$$

This is equivalent to finding y such that the following system of equations is satisfied.

$$(2) \qquad \begin{aligned} y(t) &= c_0 + \int_a^t y'(s)\, ds \\ y'(t) &= c_1 + \int_a^t y^{(2)}(s)\, ds \\ &\cdots\cdots\cdots\cdots\cdots \\ y^{(n-1)}(t) &= c_{n-1} + \int_a^t y^{(n)}(s)\, ds = c_{n-1} + \int_a^t \sum_{k=0}^{n-1} -a_k(s) y^{(k)}(s) \\ &\qquad + f(s)\, ds \end{aligned}$$

Letting Y, C and F be the $1 \times n$ matrices,

$$Y(t) = \begin{bmatrix} y(t) \\ y'(t) \\ \cdots \\ y^{(n-1)}(t) \end{bmatrix} \qquad C = \begin{bmatrix} c_0 \\ c_1 \\ \cdots \\ c_{n-1} \end{bmatrix} \qquad F(t) = \begin{bmatrix} 0 \\ 0 \\ \cdots \\ f(t) \end{bmatrix}$$

and letting $A(t)$ be the $n \times n$ matrix (defined for almost every $t \in I$),

$$A(t) \begin{bmatrix} 0 & 1 & 0 & \cdots & \cdots & 0 \\ 0 & 0 & 1 & 0 & \cdots & \cdots \\ \cdots & \cdots & \cdots & \cdots & \cdots & \cdots \\ 0 & 0 & \cdots & \cdots & \cdots & 1 \\ -a_0(t) & -a_1(t) & \cdots & \cdots & \cdots & -a_{n-1}(t) \end{bmatrix}$$

we may write (2) in the form

$$(3) \qquad Y(t) = C + \int_a^t A(s)Y(s) + F(s)\, ds$$

with the understanding that an integral of a matrix B is the matrix of integrals of the entries of B. The problem now is to solve (3). To do this, the method of successive approximations is used. Define

$$
\begin{aligned}
& Y_0(t) = C \\
(4) \qquad & \\
& Y_{k+1}(t) = C + \int_a^t A(s)Y_k(s) + F(s)\,ds \qquad k = 0, 1, \ldots
\end{aligned}
$$

Let

$$\|Y_k(t)\| = \sum_{i=1}^{n} \max_{a \le s \le t} |y_i^k(s)|$$

where the $y_i^k(s)$ are the entries in $Y_k(s)$. Defining

$$|A(s)| = \sum_{i,j} |b_{ij}(s)|$$

where the $b_{ij}(s)$ are the entries in $A(s)$, it follows easily from (4) that

$$(5) \quad \|Y_{k+1}(t) - Y_k(t)\| \le \int_a^t |A(s)|\,\|Y_k(s) - Y_{k-1}(s)\|\,ds \qquad k = 1, 2, \ldots$$

The integral in (5) exists, since $|A(s)|$ is integrable and $\|Y_k(s) - Y_{k-1}(s)\|$ is continuous as a consequence of the continuity of the entries of $Y_k(s) - Y_{k-1}(s)$. It will now be shown, by induction, that for all $t \in I$,

$$(6) \qquad \|Y_{k+1}(t) - Y_k(t)\| \le \frac{MJ^k(t)}{k!} \qquad 0 \le k$$

where $J(t) = \int_a^t |A(s)|\,ds$ and $M = \|Y_1(b) - Y_0(b)\|$.
When $k = 0$, (6) is trivial. Suppose (6) holds for $k = j - 1$. Since $J'(s) = |A(s)|$ a.e. and $J(a) = 0$, (5) and the induction hypothesis imply

$$
\begin{aligned}
\|Y_{j+1}(t) - Y_j(t)\| &\le \int_a^t |A(s)|\,\|Y_j(s) - Y_{j-1}(s)\|\,ds \\
&\le \frac{M}{(j-1)!} \int_a^t J'(s)J^{j-1}(s)\,ds \\
&= \frac{MJ^j(t)}{j!}
\end{aligned}
$$

Thus (6) holds by induction. Now, $\|Y_m(t) - Y_k(t)\|$ converges uniformly to zero as $m, k \to \infty$; for if $m > k$, then (6) implies

$$
\begin{aligned}
\|Y_m(t) - Y_k(t)\| &\le \|Y_{k+1}(t) - Y_k(t)\| + \|Y_{k+2}(t) - Y_{k+1}(t)\| \\
&\qquad + \cdots + \|Y_m(t) - Y_{m-1}(t)\| \\
&\le M \sum_{j=k}^{m-1} \frac{J^j(t)}{j!} \le M \sum_{j=k}^{m-1} \frac{J^j(b)}{j!}
\end{aligned}
$$

Since $e^{J(b)} = \sum_{j=0}^{\infty} \frac{J^j(b)}{j!}$, the assertion follows. Hence each entry $y_i^k(t)$ of $Y_k(t)$ converges uniformly on I as $k \to \infty$ to a continuous function $y_i(t)$. Letting $Y(t)$ be the column matrix with entries $y_i(t)$, it follows from letting $k \to \infty$ in (4) that Y satisfies (3). Consequently a solution to (1) exists.

By what was just proved, there exist $y_0, y_1, \ldots, y_{n-1}$ in $\mathfrak{N}(T)$ such that

$$y_i^{(k)}(a) = \delta_{ik} \qquad 0 \leq i, k \leq n-1, \quad \delta_{ik} \text{ the Kronecker delta}$$

The set $\{y_0, \ldots, y_{n-1}\}$ is linearly independent; for if $0 = y = \sum_{i=0}^{n-1} \alpha_i y_i$, then $0 = y^{(k)}(a) = \alpha_k$, $0 \leq k \leq n-1$. Thus $\alpha(T) = n$, since $\alpha(T) \leq n$ by the remark preceding Definition VI.2.6.

Proof of (ii). The proof uses the method of variation of constants. Suppose $f \in \mathfrak{L}_q(I) \subset \mathfrak{L}_1(I)$ is given. We seek a $y \in \mathfrak{D}(T)$ so that $Ty = f$. Let $y_1, y_2, \ldots, y_n$ be a basis for $\mathfrak{N}(T)$. Express y in the form $y(t) = \sum_{k=1}^{n} c_k(t) y_k(t)$, where the $c_k(t)$ are to be determined. The conditions which are initially put on the c_k are that they behave like constants when we differentiate y successively. That is to say, we want

$$\begin{aligned} y'(t) &= \sum_{k=1}^{n} c_k(t) y_k'(t) \\ y^{(2)}(t) &= \sum_{k=1}^{n} c_k(t) y_k^{(2)}(t) \\ &\cdots\cdots\cdots\cdots\cdots \\ y^{(n-1)}(t) &= \sum_{k=1}^{n} c_k(t) y_k^{(n-1)}(t) \end{aligned} \tag{7}$$

The system of Eqs. (7) holds, provided that when we differentiate y successively, we obtain

$$\begin{aligned} c_1'(t)y_1(t) + c_2'(t)y_2(t) + \cdots + c_n'(t)y_n(t) &= 0 \\ c_1'(t)y_1'(t) + c_2'(t)y_2'(t) + \cdots + c_n'(t)y_n'(t) &= 0 \\ \cdots\cdots\cdots\cdots\cdots\cdots\cdots\cdots\cdots \\ c_1'(t)y_1^{(n-2)}(t) + c_2'(t)y_2^{(n-2)}(t) + \cdots + c_n'(t)y_n^{(n-2)}(t) &= 0 \end{aligned} \tag{8}$$

Assuming that (7) holds and recalling that $\tau y_j = 0, 1 \leq j \leq n$, it follows in a straightforward manner, that

$$(\tau y)(t) = a_n(t)\Big(\sum_{k=1}^{n} c_k'(t)y_k^{(n-1)}(t)\Big) \text{ a.e.}$$

Thus if the c_k can be determined so that the system of Eqs. (8), together with the equation

$$(9) \quad c_1'(t)y_1^{(n-1)}(t) + c_2'(t)y_2^{(n-1)}(t) + \cdots + c_n'(t)y_n^{(n-1)}(t) = \frac{f(t)}{a_n(t)} \text{ a.e.}$$

holds, then $Ty = f$. An inspection of the proof of Theorem VI.2.5 verifies that the determinant

$$\begin{vmatrix} y_1(t) & y_2(t) & \cdots & y_n(t) \\ y_1'(t) & y_2'(t) & \cdots & y_n'(t) \\ \cdots & \cdots & \cdots & \cdots \\ y_1^{(n-1)}(t) & y_2^{(n-1)}(t) & \cdots & y_n^{(n-1)}(t) \end{vmatrix}$$

is not zero for any $t \in I$. Thus, for almost every $t \in I$, the system of Eqs. (8) and (9) can be solved for $c_k'(t)$ by Cramer's rule. Choosing $c_k(t) = \int_a^t c_k'(s)\,ds$, we obtain

$$(10) \qquad y(t) = \sum_{k=1}^{n} c_k(t)y_k(t) = \int_a^t G(s, t)\,\frac{f(s)}{a_n(s)}\,ds$$

where G is given in (*ii*).

Proof of (*iii*). It follows from (10) and Hölder's inequality that

$$(11) \qquad |y(t)| \leq \left\|\frac{1}{a_n}\right\|_\infty \Big(\int_a^t |G(s, t)|^{q'}\,ds\Big)^{1/q'} \|Ty\|_q \qquad 1 < q \leq \infty$$

and

$$(12) \qquad |y(t)| \leq \left\|\frac{1}{a_n}\right\|_\infty \max_{a \leq s \leq t} |G(s, t)|\;\|Ty\|_1 \qquad q = 1$$

Since $d(y, \mathfrak{N}(T)) \leq \|y\|_p$, the proof of (*iii*) follows from (11) and (12). Note that since G is continuous, $\max_{a \leq s \leq t} |G(s, t)|$ is continuous as a function of t and, in particular, is in $\mathfrak{L}_p(I)$, $1 \leq p \leq \infty$.

VI.3.2 Theorem. *Let I be an arbitrary interval and let τ be the differential expression*

$$\tau = \sum_{k=0}^{n} a_k D^k \qquad a_k \text{ locally integrable on } I, \quad 0 \leq k \leq n-1$$

$$\frac{1}{a_n} \text{ locally in } \mathfrak{L}_\infty(I)$$

Then the maximal operator corresponding to (τ, p, q), $1 \leq p, q \leq \infty$, is closed.

Proof. Let T be the maximal operator corresponding to (τ, p, q). Suppose

$$f_n \to f \in \mathfrak{L}_p(I)$$
(1)
$$Tf_n \to g \in \mathfrak{L}_q(I)$$

For J a compact subinterval of I, define T_J to be the maximal operator corresponding to (τ, p, q, J). Considering f_n and f as elements of $\mathfrak{L}_p(J)$ and g as an element of $\mathfrak{L}_q(J)$, it is clear from (1) that

$$f_n \to f \in \mathfrak{L}_p(J)$$
(2)
$$T_J f_n \to g \in \mathfrak{L}_q(J)$$

Now, by Theorem VI.3.1, T_J is surjective and $\gamma(T_J) > 0$. Hence T_J is closed by Theorem IV.1.7. Therefore we may conclude from (2) that $f \in \mathfrak{D}(T_J)$ and $T_J f = g$. Since J was an arbitrary compact subinterval of I and f is in $\mathfrak{L}_p(I)$, it follows that $f \in \mathfrak{D}(T)$ and $Tf = g$. Thus T is closed.

VI.3.3 Corollary. *Let $I = [a, b]$ be compact and let τ be as in Theorem* VI.3.1. *Suppose T is a* 1-1 *closed operator which is a restriction of the maximal operator corresponding to (τ, p, q), $1 \leq p \leq \infty$, $1 < q \leq \infty$. Then T^{-1} is compact.*

Proof. Suppose $\{f_k\}$ is a bounded sequence in $\mathfrak{R}(T) \subset \mathfrak{L}_q(I)$. By Theorem VI.3.1, there exists a G, continuous on $I \times I$, and a sequence $\{z_k\}$ in $\mathfrak{N}(T_1)$, where T_1 is the maximal operator corresponding to (τ, p, q), such that

$$(1) \qquad (T^{-1}f_k)(t) - z_k(t) = y_k(t) = \int_a^t G(s, t) \frac{f_k(s)}{a_n(s)} \, ds$$

Now Hölder's inequality implies

$$|y_k(t)| \leq M \|f_k\|_1 \leq M(b - a)^{1/q'} \|f_k\|_q$$

where $M = \|1/a_n\|_\infty \sup_{a \leq s,t \leq b} |G(s, t)|$. Hence $\{y_k\}$ is uniformly bounded on I and, in particular, is bounded in $\mathfrak{L}_p(I)$. Moreover, $\{z_k\}$ is bounded in $\mathfrak{L}_p(I)$. If this is not the case, then there exists a subsequence $\{z_{k'}\}$ such that for $x_{k'} = T^{-1}f_{k'}$

$$(2) \qquad \frac{(x_{k'} - z_{k'})}{\|z_{k'}\|} \to 0 \qquad \text{and} \qquad \frac{Tx_{k'}}{\|z_{k'}\|} \to 0$$

Since $\mathfrak{N}(T_1)$ is finite-dimensional, there exists a subsequence $\{z_{k''}\}$ of $\{z_{k'}\}$ such that $z_{k''}/\|z_{k''}\| \to v \in \mathfrak{N}(T_1)$. Thus, by (2)

$$\frac{x_{k''}}{\|z_{k''}\|} \to v \qquad \text{and} \qquad \frac{Tx_{k''}}{\|z_{k''}\|} \to 0$$

Hence $v \in \mathfrak{D}(T)$ and $Tv = 0$, which is impossible since T is 1-1 and $\|v\| = 1$. Therefore $\{z_k\}$ is bounded. Consequently $\{z_k\}$ has a convergent subsequence. Hence in order to prove that T^{-1} is compact, it suffices to prove that $\{y_{k'}\}$ has a convergent subsequence, where $\{y_{k'}\}$ is any subsequence of $\{y_k\}$. For convenience, take $\{y_{k'}\}$ to be $\{y_k\}$. We show that $\{y_k\}$ is equicontinuous. Let $\varepsilon > 0$ be given. Since G is uniformly continuous on $I \times I$, there exists a $\delta = \delta(\epsilon) > 0$ such that for all $s \in I$,

$$(3) \qquad |G(s, t_1) - G(s, t_2)| \leq \varepsilon \qquad |t_1 - t_2| < \delta$$

Now

$$y_k(t_2) - y_k(t_1) = \int_a^{t_2} G(s, t_2) \frac{f_k(s)}{a_n(s)}\, ds - \int_a^{t_1} G(s, t_2) \frac{f_k(s)}{a_n(s)}\, ds + \int_a^{t_1} [G(s, t_2) - G(s, t_1)] \frac{f_k(s)}{a_n(s)}\, ds$$

Thus (3) and Hölder's inequality imply

$$|y_k(t_2) - y_k(t_1)| \leq M|t_1 - t_2|^{1/q'}\|f_k\|_q + \varepsilon \left\|\frac{1}{a_n}\right\|_\infty (b - a)^{1/q'}\|f_k\|_q$$

Since $\{\|f_k\|_q\}$ is bounded, it follows that $\{y_k\}$ is equicontinuous. Therefore, by the Ascoli-Arzelà theorem, $\{y_k\}$ has a subsequence converging in $C(I)$ which, in turn, must converge in $\mathfrak{L}_p(I)$. Hence T^{-1} is compact.

The following example exhibits very simple maximal operators which do not have closed range.

VI.3.4 Example. Let $I = [0, \infty)$ and let $\tau_\lambda = D - \lambda e^{-it}$, where λ is a scalar. We show that for $T_\lambda = T_{\tau_\lambda, p, q}$, (p, q) admissible, and all λ, T_λ

does not have a closed range, or equivalently, by Theorem VI.2.11, T_λ is not surjective.

Given f locally integrable on I, the general solution in $A_1(I)$ for which $\tau_\lambda y(t) = y'(t) - \lambda e^{-it}y(t) = f(t)$ a.e. is given by

$$y(t) = ce^{-\varphi(t)} + e^{-\varphi(t)} \int_0^t e^{\varphi(s)} f(s)\, ds \tag{1}$$

where $\varphi(t) = -i\lambda e^{-it}$ and c is an arbitrary constant.

Since $|\varphi(t)| = |\lambda|$,

$$e^{|\lambda|} \geq |e^{-\varphi(t)}| \geq e^{-|\varphi(t)|} = e^{-|\lambda|} \qquad t \in I \tag{2}$$

i. $q = \infty$. Let $f(s) = e^{-\varphi(s)}$, $s \in I$. Then by (2), $e^{|\lambda|} \geq |f(s)|$, which means that f is in $\mathfrak{L}_\infty(I)$. Then from (1), the general solution in $A_1(I)$ to $\tau y = f$ a.e. is

$$y(t) = ce^{-\varphi(t)} + e^{-\varphi(t)} \int_0^t 1\, ds = (c + t)e^{-\varphi(t)}$$

Hence $|y(t)| \geq |c + t|e^{-|\lambda|}$, which is not in $\mathfrak{L}_p(I)$, $1 \leq p \leq \infty$. Thus f is not in $\mathfrak{R}(T_\lambda)$ when $1 \leq p \leq \infty$ and $q = \infty$.

ii. $1 < q < \infty$. Define f on I by

$$f(t) = \frac{e^{-\varphi(t)}}{k} \qquad k - 1 \leq t < k, \quad k = 1, 2, \ldots$$

Then f is in $\mathfrak{L}_q(I)$, since $|e^{-\varphi(t)}| \leq e^{|\lambda|}$ and $\sum_{k=1}^{\infty} 1/k^q < \infty$. Let y be as in (1). Given a positive integer N, it follows from (2) and the definition of f that for $N + 1 \geq t \geq N$,

$$|y(t)| = \left| ce^{-\varphi(t)} + e^{-\varphi(t)} \left(\sum_{k=1}^{N} \frac{1}{k} + \int_N^t e^{\varphi(s)} f(s)\, ds \right) \right| \geq e^{-|\lambda|} \left| c + \sum_{k=1}^{N} \frac{1}{k} \right| - 1$$

Since $\sum_{k=1}^{\infty} 1/k = \infty$, y is not in $\mathfrak{L}_p(I)$, $1 \leq p \leq \infty$. Hence f is not in $\mathfrak{R}(T_{\tau\lambda,p,q})$ when $1 \leq p \leq \infty$ and $1 < q < \infty$.

To summarize, it has been shown that for all λ

(3) $T_{\tau\lambda,p,q}$ is not surjective when $1 \leq p \leq \infty$ and $1 < q \leq \infty$.

The last case to consider for (p, q) admissible is

iii. $q = 1$, $1 \leq p < \infty$. Since $q' = \infty$, $1 < p' \leq \infty$, and $(\tau_\lambda)^* = -\tau_{-\lambda}$, it follows from (3) that $T_{(\tau\lambda)^*,q',p'}$ is not surjective and therefore, by Theorem VI.2.11, $T_{\tau\lambda,p,q}$ is not surjective.

The above example also shows that the essential resolvent:

$$\{\lambda \mid \lambda I - T_{\nu,p,p} \text{ has closed range}\}$$

is empty when $\nu = e^{it}D$ and $1 \leq p \leq \infty$. This is a consequence of the observation that $T_{\nu,p,p} - \lambda I = e^{it}T_{\tau\lambda,p,p}$.

VI.4 EXTENSION THEOREMS

It is seen from Theorems VI.3.1 and VI.2.11 that minimal operators determined by certain differential expressions on a compact interval have a bounded inverse, while the maximal operators are surjective. This section treats the following problem.

Let T_1 be a linear operator mapping a subspace of a Banach space onto a Banach space and let T_0 be a restriction of T_1 such that T_0 has a bounded inverse. Under what conditions does there exist a linear operator T which is a restriction of T_1 and an extension of T_0, written $T_0 \subset T \subset T_1$, such that T is surjective and has a bounded inverse?

The problem was considered by Višk [1] in his investigation of partial differential operators in a Hilbert space setting and by Browder [2] for more general spaces.

VI.4.1 Definition. *Let T_1 be a linear operator from one normed linear space into another and let T_0 be a restriction of T_1. The pair (T_0, T_1) is called* **potentially solvable** *if T_1 is surjective and T_0 has a bounded inverse.*

The pair of operators (T_0, T_1) is called **solvable** *if there exists a linear operator T such that $T_0 \subset T \subset T_1$, T is surjective, and T has a bounded inverse. We shall also say that T* **solves** *(T_0, T_1).*

VI.4.2 Theorem. *Let X and Y be Banach spaces. Suppose that (T_0, T_1) is potentially solvable with $\mathfrak{D}(T_1) \subset X$ and $\mathfrak{R}(T_1) = Y$. A sufficient condition that (T_0, T_1) be solvable is that T_1 be closed and there exist projections from X onto $\mathfrak{N}(T_1)$ and Y onto $\mathfrak{R}(T_0)$.*

An operator T which solves (T_0, T_1) may be constructed as follows. Suppose

$$X = X_0 \oplus \mathfrak{N}(T_1) \qquad Y = Y_0 \oplus \mathfrak{R}(T_0)$$

where X_0 and Y_0 are closed subspaces. Define T by

$$\mathfrak{D}(T) = \mathfrak{D}(T_0) \oplus X_0 \cap T_1^{-1}Y_0$$

$$T(u + v) = T_0u + T_1v \qquad u \in \mathfrak{D}(T_0),\ v \in X_0 \cap T_1^{-1}Y_0$$

Proof. In a very natural way, an operator T which solves (T_0, T_1) will be determined after considering some properties T must have.

Now,

$$Y = \mathcal{R}(T_0) \oplus Y_0 \tag{1}$$

where Y_0 is the closed subspace which is the kernel of a projection onto $\mathcal{R}(T_0)$. Thus if T were to exist, then for each x in $\mathfrak{D}(T)$ there exists a u in $\mathfrak{D}(T_0)$ and a w in Y_0 such that $Tx = T_0u + w$. Since T is to be 1-1 and surjective,

$$x = u + T^{-1}w \in \mathfrak{D}(T_0) + T_1^{-1}Y_0$$

(T_1^{-1} is used in the set theoretic sense.) We note that

$$\mathfrak{D}(T_0) \cap T_1^{-1}Y_0 = (0)$$

for if z is in $\mathfrak{D}(T_0)$ and T_1z is in Y_0, then $T_0z = T_1z \in \mathcal{R}(T_0) \cap Y_0 = (0)$. Therefore $z = 0$ since T_0 is 1-1. Hence we must have

$$\mathfrak{D}(T_0) \subset \mathfrak{D}(T) \subset \mathfrak{D}(T_0) \oplus T_1^{-1}Y_0$$

and on $\mathfrak{D}(T)$,

$$T(u + v) = T_0u + T_1v \qquad u \in \mathfrak{D}(T_0),\, v \in T_1^{-1}Y_0$$

If $\mathfrak{D}(T)$ is chosen to be $\mathfrak{D}(T_0) \oplus T_1^{-1}Y_0$, then T might not be 1-1; for if $v \neq 0$ is in $\mathfrak{N}(T_1)$, then v is in $T_1^{-1}Y_0$. Thus $Tv = T_1v = 0$. To avoid this situation, we make use of the hypothesis that

$$X = \mathfrak{N}(T_1) \oplus X_0 \tag{2}$$

where X_0 is a closed subspace of X by defining T by

$$\mathfrak{D}(T) = \mathfrak{D}(T_0) \oplus X_0 \cap T_1^{-1}Y_0 \tag{3}$$

$$T(u + v) = T_0u + T_1v \qquad u \in \mathfrak{D}(T_0),\, v \in X_0 \cap T_1^{-1}Y_0 \tag{4}$$

From the way T was determined it is clear that $T_0 \subset T \subset T_1$ and that T is 1-1.

We now show that T is surjective. Suppose $y \in Y$. By (1), there exist $u \in \mathfrak{D}(T_0)$ and $w \in Y_0$ such that $y = T_0u + w$. Since T_1 is surjective, there exists a $v \in \mathfrak{D}(T_1)$ such that $T_1v = w$. Thus v is in $T_1^{-1}Y_0$. By (2), there exists a $v_0 \in \mathfrak{N}(T_1) \subset T_1^{-1}Y_0$ such that $v - v_0$ is in X_0. Hence $v - v_0$ is in $X_0 \cap T_1^{-1}Y_0$, and $u + (v - v_0)$ is in $\mathfrak{D}(T)$. Furthermore,

$$T(u + (v - v_0)) = T_0u + w = y$$

whence T is surjective.

To prove that T^{-1} is continuous, it suffices, by the closed-graph theorem, to prove that T is closed. Suppose

$$(5)\qquad \begin{gathered} u_n + v_n \to x \qquad u_n \in \mathfrak{D}(T_0), \quad v_n \in X_0 \cap T_1^{-1}Y_0 \\ T(u_n + v_n) = T_0u_n + T_1v_n \to y \end{gathered}$$

Since T_1 is a closed extension of T, $y = T_1x$. The existence of a projection P from Y onto $\mathfrak{R}(T_0)$ with $\mathfrak{N}(P) = Y_0$, together with (5), implies

$$(6)\qquad T_0u_n \to w = Py \in \mathfrak{R}(T_0)$$

Since T_0^{-1} is continuous,

$$(7)\qquad u_n \to T_0^{-1}w$$

Hence from (5), (6), and (7),

$$(8)\qquad \begin{gathered} v_n \to x - T_0^{-1}w \\ T_1v_n \to y - w = T_1x - w \end{gathered}$$

But v_n is in $X_0 \cap T_1^{-1}Y_0$, and both X_0 and Y_0 are closed. Therefore, by (8), $x - T_0^{-1}w$ is in X_0 and $T_1x - w$ is in Y_0. Since

$$T_1(x - T_0^{-1}w) = T_1x - w$$

we have

$$x - T_0^{-1}w \in X_0 \cap T_1^{-1}Y_0$$

Hence x is in $\mathfrak{D}(T_0) \oplus X_0 \cap T_1^{-1}Y_0 = \mathfrak{D}(T)$, and $Tx = T_1x = y$. Thus T is closed and the theorem is proved.

VI.4.3 Corollary. *Suppose X and Y are Hilbert spaces. If (T_0, T_1) is potentially solvable and T_1 is closed, then (T_0, T_1) is solvable.*

Proof. Since T_0 has a bounded inverse, the minimal closed extension $\bar{T}_0$ also has a bounded inverse. Hence $\mathfrak{R}(\bar{T}_0)$ is closed. Since X and Y can be projected onto $\mathfrak{N}(T_1)$ and $\mathfrak{R}(\bar{T}_0)$, respectively, the corollary follows from Theorem VI.4.2.

VI.4.4 Corollary. Let $\tau = \sum_{k=0}^{n} a_kD^k$, $a_k \in C^k(I)$, $0 \le k \le n$, $a_n(t) \neq 0$, *$t \in I$, and let (p, q) be admissible. If $(T_{0,\tau,p,q}, T_{\tau,p,q})$ is potentially solvable, then the pair is solvable.*

Proof. Theorems VI.2.7, II.1.16 and VI.4.2.

For the remainder of this chapter, τ is the differential expression

$$\tau = \sum_{k=0}^{n} a_k D^k \qquad a_k \in C^k(I), \quad 0 \leq k \leq n$$

$$a_n(t) \neq 0, \quad t \in I$$

and (p, q) is admissible.

VI.4.5 Corollary. *If I is compact, then $(T_{o,\tau,p,q}, T_{\tau,p,q})$ is solvable. If $1 < p \leq \infty$ and $1 < q \leq \infty$, any linear operator which solves $(T_{o,\tau,p,q}, T_{\tau,p,q})$ has a compact inverse.*

Proof. Theorems VI.3.1 and VI.2.11 and Corollaries VI.4.4 and VI.3.3.

The construction of T in Theorem VI.4.2 leads to the following characterization of all the differential operators which solve a pair of minimal and maximal operators.

VI.4.6 Theorem. *Suppose (T_0, T_1) is solvable, where $T_0 = T_{o,\tau,p,q}$ and $T_1 = T_{\tau,p,q}$. Then an operator T which solves the pair can be constructed as follows.*

i. Let $v_1, v_2, \ldots, v_m$ be a basis for $\mathfrak{N}(T_{\tau^*,q',p'})$ and let $y_1, y_2, \ldots, y_m$ be in $\mathfrak{L}_q(I)$ such that

$$\int_I v_i y_j = \delta_{ij} \qquad i, j = 1, 2, \ldots, m; \quad \delta_{ij} \text{ the Kronecker delta}$$

ii. Let $x_1, x_2, \ldots, x_m$ be elements in $\mathfrak{D}(T_1)$ such that

$$T_1 x_i = y_i \qquad 1 \leq i \leq m$$

If $\{z_1, z_2, \ldots, z_m\}$ is a set of elements in $\mathfrak{N}(T_1)$ (the elements need not be distinct), define T as follows.

$$\mathfrak{D}(T) = \mathfrak{D}(T_0) \oplus sp\ \{x_i + z_i\}$$

$$T(u + v) = T_0 u + T_1 v \qquad u \in \mathfrak{D}(T_0),\ v \in \text{sp}\ \{x_i + z_i\}$$

Then T solves (T_0, T_1).

Conversely, any T which solves (T_0, T_1) is of the form given by (ii), where $\{y_1, y_2, \ldots, y_m\}$ is linearly independent and

$$Y = \mathfrak{R}(T_0) \oplus \text{sp}\ \{y_1, y_2, \ldots, y_m\}$$

Proof. Condition (*i*) is used to obtain a Y_0 as in Theorem VI.4.2, while (*ii*) is used for the purpose of exhibiting $X_0 \cap T_1^{-1}Y_0$.

Writing $T_* = T_{T^*,q',p'}$, we note that $\mathcal{R}(T_*)$ is closed by Theorem VI.2.7. Hence Theorems IV.1.2 and VI.2.3 show that

$$\mathcal{R}(T_0) = {}^{\perp}\mathfrak{N}(T_0') = {}^{\perp}\mathfrak{N}(T_*) \qquad 1 \leq p, q < \infty$$

$$\mathcal{R}(T_0) = \mathcal{R}(T_*') = \mathfrak{N}(T_*)^{\perp} \qquad 1 < p, q \leq \infty$$

Thus if $\{v_1, \ldots, v_m\}$ and $\{y_1, \ldots, y_m\}$ are as in (*i*), then

$$Y_0 = \text{sp}\ \{y_1, \ldots, y_m\}$$

and $\mathcal{R}(T_0)$ are linearly independent. Since

$$\beta(T_0) = \alpha(T_0') = \alpha(T_*) \qquad 1 \leq p, q < \infty$$

$$\beta(T_0) = \beta(T_*') = \alpha(T_*) \qquad 1 < p, q \leq \infty$$

it follows that $Y = \mathcal{R}(T_0) \oplus Y_0$. Let both $\{x_1, x_2, \ldots, x_m\}$ and $\{z_1, \ldots, z_m\}$ be as in (*ii*). Then sp $\{x_i + z_i\}$ is a subspace of $T_1^{-1}Y_0$. We obtain next an X_0 with the properties required in Theorem VI.4.2. Let W be a closed subspace of X such that

$$X = \mathfrak{N}(T_1) \oplus \text{sp}\ \{x_i + z_i\} \oplus W \tag{1}$$

The existence of W is assured since $\mathfrak{N}(T_1) + \text{sp}\ \{x_i + z_i\}$ is finite-dimensional. Define $X_0 = \text{sp}\ \{x_i + z_i\} \oplus W$. Since

$$T_1^{-1}Y_0 = \text{sp}\ \{x_i + z_i\} \oplus \mathfrak{N}(T_1)$$

it follows from (1) that

$$X_0 \cap T_1^{-1}Y_0 = \text{sp}\ \{x_i + z_i\}$$

Hence the operator T solves (T_0, T_1) by Theorem VI.4.2.

Conversely, suppose T solves (T_0, T_1), and $Y = \mathcal{R}(T_0) \oplus Y_0$, $Y_0 = \text{sp}\ \{y_1, y_2, \ldots, y_m\}$, $\dim Y_0 = m$. Then $\mathfrak{D}(T) \subset \mathfrak{D}(T_0) \oplus T_1^{-1}Y_0$, as was shown in the proof of Theorem VI.4.2. Since T is surjective, we are assured of elements $u_1, u_2, \ldots, u_m$ in $\mathfrak{D}(T)$ such that $Tu_i = T_1x_i = y_i$, $1 \leq i \leq m$. Thus there exist $z_1, z_2, \ldots, z_m$ in $\mathfrak{N}(T_1)$ such that $u_i = x_i + z_i$. Given $x = u + v$ in $\mathfrak{D}(T)$ with $u \in \mathfrak{D}(T_0)$ and

$v \in T_1{}^1 Y_0$, we have that v is in $\mathfrak{D}(T)$ and

$$Tv = \sum_{i=1}^{m} \alpha_i y_i = \sum_{i=1}^{m} \alpha_i T u_i = T \sum_{i=1}^{m} \alpha_i (x_i + z_i)$$

Since T is 1-1, $v = \sum_{i=1}^{m} \alpha_i (x_i + z_i)$. Thus T has the form given in (*ii*).

VI.5 BOUNDARY VALUE FUNCTIONALS

VI.5.1 Definition. *A closed linear operator which is an extension of the minimal operator and a restriction of the maximal operator corresponding to* (τ, p, q) *is called a* ***differential operator*** *corresponding to* (τ, p, q).

Writing $T_1 = T_{\tau,p,q}$ and $T_0 = T_{o,\tau,p,q}$, we let D_1 be the space $\mathfrak{D}(T_1)$ with the T_1-norm $\| \quad \|_1$ defined by

$$\|x\|_1 = \|x\| + \|T_1 x\|$$

Since T_1 is closed, D_1 is a Banach space. Suppose T is a differential operator corresponding to (τ, p, q). If $D_0 = \mathfrak{D}(T_0)$ and $D_T = \mathfrak{D}(T)$ are equipped with the T_1-norm, then D_0 and D_T are closed subspaces of D_1 with $D_0 \subset D_T \subset D_1$. On the other hand, given a closed subspace M of D_1 such that $D_0 \subset M$, the operator T_M which is defined to be T_1 restricted to M is a differential operator. Thus there is a 1-1 correspondence between the differential operators and the closed subspaces M of D_1 which contain D_0, the correspondence being $M \rightarrow T_M$.

Suppose M is a closed subspace of D_1 and contains D_0. Then by Theorem II.3.4, $M = {}^{\perp}(M^{\perp})$, where $M^{\perp} \subset D_1'$, the conjugate space of D_1. Since D_0 is contained in M, it is obvious that $M^{\perp} \subset D_0{}^{\perp}$. Thus M is the intersection of the null manifolds of a set of linear functionals which are continuous on D_1 and which annihilate D_0, the set of functionals being $M^{\perp}$.

Conversely, given any set $\{B_\alpha\}$ of linear functionals which are continuous on D_1 and which annihilate D_0, the space $M = \bigcap_\alpha \mathfrak{N}(B_\alpha)$ is a closed subspace of D_1 which contains D_0.

The discussion leads to the following definition and theorem.

VI.5.2 Definition. *Let* D_1 *and* D_0 *be as in the above discussion. A functional* B *which is continuous on* D_1 *and which annihilates* D_0, *that is,* $B \in D_0{}^{\perp} \subset D_1'$, *is called a* ***boundary value functional.***

VI.5.3 Theorem. *There is a 1-1 correspondence between the differential operators corresponding to (τ, p, q) and the set of subspaces of the form*

$$\bigcap_{\alpha} \mathfrak{N}(B_\alpha) \tag{1}$$

where $\{B_\alpha\}$ is a set of boundary value functionals. The differential operators are obtained by restricting T_1 to sets of the form (1); i.e., differential operator T is the restriction of T_1 to those $y \in \mathfrak{D}(T_1)$ which satisfy the equations

$$B_\alpha y = 0 \qquad \text{all } \alpha$$

VI.5.4 Definition. *If T is the restriction of the maximal operator to $\bigcap_{\alpha} \mathfrak{N}(B_\alpha)$, where $\{B_\alpha\}$ is a set of boundary value functionals, T is said to be* ***determined*** *by $\{B_\alpha\}$.*

VI.5.5 Example. The notation used in this example will be that given in the discussion preceding Definition VI.5.2.

Suppose that I contains one of its endpoints a. Let functionals B_k be defined on D_1 by

$$B_k(y) = \sum_{j=0}^{n-1} b_{kj} y^{(j)}(a) \qquad k = 1, 2, \ldots, n$$

where the b_{kj} are constants. Then each B_k is a boundary value functional. To see this, let I_0 be any compact subinterval of I which contains a as an endpoint. Define D_2 to be the space $\mathfrak{D}(T_1)$ with norm defined by

$$\|y\|_2 = \|y\|_1 + \sum_{j=0}^{n-1} \max_{t \in I_0} |y^{(j)}(t)|$$

It is not difficult to verify that D_2 is a Banach space. Since $\|y\|_1 \leq \|y\|_2$ for all $y \in D_1$, we have that D_1 and D_2 are isomorphic by the open-mapping principle. Each B_k is clearly bounded on D_2 and therefore bounded on D_1. By Lemma VI.2.9, $y \in \mathfrak{D}(T_0)$ implies $y^{(j)}(a) = 0$, $0 \leq j \leq n - 1$. Hence $B_k D_0 = 0$, $1 \leq k \leq n$. Thus $\{B_k\}$ is a set of boundary value functionals, and the restriction of the maximal operator to those $y \in \mathfrak{D}(T_1)$ which satisfy the "boundary conditions"

$$B_k y = \sum_{j=0}^{n-1} b_{kj} y^{(j)}(a) = 0 \qquad k = 1, 2, \ldots, n$$

is a differential operator. If the determinant of the matrix (b_{kj}) is not zero, it follows from Lemma VI.2.4 that the differential operator is 1-1.

Assuming that the minimal operator has a closed range, the next theorem characterizes, in terms of boundary value functionals, the differential operators.

VI.5.6 Theorem. *Let $T_{0,\tau,p,q}$ have a closed range and let T be a differential operator corresponding to (τ, p, q). If $\{B_\alpha\}$ is a set of boundary value functionals which determine T, then*

$$\dim \operatorname{sp} \{B_\alpha\} = \kappa(T_1) - \kappa(T) = \dim \frac{\mathfrak{D}(T_1)}{\mathfrak{D}(T)} < \infty$$

Thus T is the restriction of the maximal operator to those $y \in \mathfrak{D}(T_1)$ which satisfy the conditions

$$0 = B_1 y = B_2 y = \cdots = B_m y$$

where $m = \kappa(T_1) - \kappa(T)$ and $\{B_1, B_2, \ldots, B_m\}$ is any linearly independent subset of $\{B_\alpha\}$.

Proof. By Theorem VI.2.7, $T_0 = T_{0,\tau,p,q}$ and $T_1 = T_{\tau,p,q}$ are Fredholm operators. Since $\operatorname{sp}\{B_\alpha\} = D_T{}^\perp \subset D_1'$, where D_T and D_1 are $\mathfrak{D}(T)$ and $\mathfrak{D}(T_1)$, respectively, with the T_1-norm, Theorem VI.2.7 implies

$$\dim \operatorname{sp} \{B_\alpha\} = \dim D_T{}^\perp = \dim \left(\frac{D_1}{D_T}\right)' = \dim \frac{D_1}{D_T} \leq \dim \frac{D_1}{\mathfrak{D}(T_0)} < \infty$$

Thus $\dim \operatorname{sp} \{B_\alpha\} = \kappa(T_1) - \kappa(T)$ by Lemma V.1.5.

VI.5.7 Corollary. *Suppose the minimal operator $T_{0,\tau,p,q}$ has a closed range. If $\{B_\alpha\}$ is a set of boundary value functionals which determines $T_{0,\tau,p,q}$ and $\{V_\beta\}$ is a set of boundary value functionals which determines $T_{0,\tau^*,q',p'}$, then*

$$\dim \operatorname{sp} \{B_\alpha\} = \dim \operatorname{sp} \{V_\beta\} = \kappa(T_{\tau,p,q}) - \kappa(T_{0,\tau,p,q}) < \infty$$

Proof. We write $T_1 = T_{\tau,p,q}$, $T_* = T_{\tau^*,q',p'}$, $T_o = T_{0,\tau,p,q}$, and $T_{*o} = T_{0,\tau^*,q',p'}$. Let $\mathfrak{B} = \operatorname{sp}\{B_\alpha\}$ and $\mathfrak{V} = \operatorname{sp}\{V_\beta\}$.

If $1 \leq p, q < \infty$, then $T_1' = T_{*o}$ and $T_o' = T_*$. Hence it follows from Theorems IV.2.3 and VI.5.6 that

$$\infty > \dim \mathfrak{B} = \kappa(T_1) - \kappa(T_o) = -\kappa(T_1') + \kappa(T_o') = \kappa(T_*) - \kappa(T_{*o}) = \dim \mathfrak{V}$$

If $1 < p, q \leq \infty$, then $T_{*o}' = T_1$ and $T_*' = T_o$. Hence

$$\infty > \dim \mathfrak{B} = \kappa(T_1) - \kappa(T_o) = \kappa(T_{*o}') - \kappa(T_*') = \kappa(T_*) - \kappa(T_{*o}) = \dim \mathfrak{V}$$

The differential operators which solve the pair consisting of the minimal and maximal operators are characterized in terms of boundary value functionals as follows.

VI.5.8 Theorem. *Suppose* $(T_{o,\tau,p,q}, T_{\tau,p,q})$ *is solvable. Let* $\{B_\alpha\}$ *be a set of boundary value functionals which determines the differential operator* T. *Then* T *solves* $(T_{o,\tau,p,q}, T_{\tau,p,q})$ *if and only if the following two conditions are satisfied.*

i. The dimension of the space $\mathcal{B}$ *spanned by the set* $\{B_\alpha\}$ *is* $\dim \mathfrak{N}(T_{\tau,p,q})$.

ii. If $y_1, y_2, \ldots, y_m$ *is a basis for* $\mathfrak{N}(T_{\tau,p,q})$ *and* $B_1, B_2, \ldots, B_m$ *is a basis for* $\mathcal{B}$, *then the determinant of the matrix* $(B_i y_j)$ *is not zero.*

Proof. Suppose T solves (T_0, T_1), where $T_0 = T_{o,\tau,p,q}$ and $T_1 = T_{\tau,p,q}$. Then by Theorem VI.5.6

$$\dim \mathcal{B} = \kappa(T_1) - \kappa(T) = \kappa(T_1) = \alpha(T_1)$$

To prove (*ii*), assume the determinant $\|(B_i y_j)\|$ is zero. Then there exist scalars $c_1, c_2, \ldots, c_m$, not all zero, such that

$$c_1 B_1 y_1 + c_2 B_1 y_2 + \cdots + c_m B_1 y_m = 0$$

$$c_1 B_2 y_1 + c_2 B_2 y_2 + \cdots + c_m B_2 y_m = 0$$

$$\cdots\cdots\cdots\cdots\cdots\cdots\cdots\cdots$$

$$c_1 B_m y_1 + c_2 B_m y_2 + \cdots + c_m B_m y_m = 0$$

Let $y = \sum_{i=1}^{m} c_i y_i$. Then $0 \neq y \in \mathfrak{N}(T_1)$ and $B_k y = 0, 1 \leq k \leq m$. Hence, by definition, y is in $\mathfrak{D}(T)$ and $0 = T_1 y = Ty$, which is impossible, since T is 1-1.

Conversely, assume (*i*) and (*ii*). Suppose $0 = Ty = T_1 y$. Then $y = \sum_{i=1}^{m} b_i y_i$ for some $b_1, b_2, \ldots, b_m$. Since y is in $\mathfrak{D}(T)$ and $\{B_i\}$ determines T,

$$0 = B_k y = \sum_{i=1}^{m} b_i B_k y_i \qquad k = 1, 2, \ldots, m \tag{1}$$

The assumption that $\|(B_i y_j)\| \neq 0$, together with (1), implies $b_i = 0$, $1 \leq i \leq m$. Thus $y = 0$, whence T is 1-1. Now, by (*ii*) and Theorem VI.5.6,

$$\infty > \alpha(T_1) = \dim \mathcal{B} = \kappa(T_1) - \kappa(T) \tag{2}$$

Since T_1 is surjective and T is 1-1, it follows from (2) that $0 = \beta(T)$; that is, T is surjective. Furthermore, the closed-graph theorem applied to T^{-1} proves that T^{-1} is continuous. Hence T solves (T_0, T_1).

For further properties of boundary value functionals, the reader is referred to Rota [1], Lemmas 5.3 to 5.5.

Kemp [1] considers the effect on the minimal and maximal operators corresponding to (τ, p, p), $1 \leq p \leq \infty$, when the leading coefficient a_n of τ vanishes in the interior of I. The conjugates of the operators are determined, and the domains of differential operators are described in terms of boundary value functionals.

VI.6 SOME A PRIORI ESTIMATES

In this section, some inequalities, called *a priori estimates*, which are used to determine properties of certain differential operators, are derived. As we shall see in the next chapter, these estimates have their counterparts in the theory of partial differential operators.

VI.6.1 Lemma. *For n a positive integer and $1 \leq p \leq \infty$, let $W^{n,p}(I)$ be the linear manifold defined by*

$$W^{n,p}(I) = \{g \mid g \in A_n(I) \cap \mathfrak{L}_p(I) \qquad g^{(n)} \in \mathfrak{L}_p(I)\}$$

i. If f is in $W^{n,p}(I)$, then all the derivatives $f^{(k)}$, $0 \leq k \leq n$, are in $\mathfrak{L}_p(I)$.

ii. For each $\varepsilon > 0$, there exists a constant K depending only on ε, p, and the length of I (I may be unbounded), such that for all $f \in W^{n,p}(I)$,

$$\|f^{(k)}\|_{p,I}^p \leq K\|f\|_{p,I}^p + \varepsilon\|f^{(n)}\|_{p,I}^p \qquad 0 \leq k \leq n-1,\ 1 \leq p < \infty$$

$$\|f^{(k)}\|_{\infty,I} \leq K\|f\|_{\infty,I} + \varepsilon\|f^{(n)}\|_{\infty} \qquad 0 \leq k \leq n-1,\ p = \infty$$

where $\|\ \|_{p,I}$ denotes the norm on $\mathfrak{L}_p(I)$.

iii. Considered as a function of the length of I, K may be chosen to be nonincreasing.

Proof. It suffices to assume that g is real-valued. The lemma will first be proved for $n = 2$. For $n = 1$, the lemma is trivial. Suppose f is in $W^{2,p}(I)$, $1 \leq p \leq \infty$. Let $\{I_j\}$ be a sequence of nonoverlapping bounded intervals of equal length L whose union is I. Let J be any I_j. Subdivide J into three subintervals of equal length and choose J_1 and J_2 to be the two subintervals which are separated by the middle third subinterval.

For $x \in J_1$ and $y \in J_2$ there exists, by the mean-value theorem, a $\xi = \xi_{x,y}$ in J such that

$$\frac{f(x) - f(y)}{x - y} = f'(\xi) \tag{1}$$

Since $|x - y| \geq L/3$, it follows from (1) and the absolute continuity of f' that for each $t \in J$,

$$|f'(t)| = \left| f'(\xi) + \int_\xi^t f''(s)\, ds \right| \leq 3L^{-1}(|f(x)| + |f(y)|) + \left| \int_J f''(s)\, ds \right| \tag{2}$$

Thus, for $p = \infty$ and $t \in J$,

$$|f'(t)| \leq 6L^{-1}\|f\|_{\infty,I} + L\|f''\|_{\infty,I}$$

Since J was an arbitrary I_j, f' is in $\mathfrak{L}_\infty(I)$ and

$$\|f'\|_{\infty,I} \leq 6L^{-1}\|f\|_{\infty,I} + L\|f''\|_{\infty,I} \tag{3}$$

Suppose $1 \leq p < \infty$. Integrating both sides of (2), first with respect to x on J_1 and then with respect to y on J_2, gives

$$\begin{aligned} 3^{-2}L^2|f'(t)| &\leq \int_{J_1} |f(x)|\, dx + \int_{J_2} |f(y)|\, dy + 3^{-2}L^2 \left| \int_J f''(s)\, ds \right| \\ &\leq \int_J |f(s)|\, ds + 3^{-2}L^2 \int_J |f''(s)|\, ds \end{aligned}$$

Thus, from Hölder's inequality,

$$\begin{aligned} |f'(t)| &\leq 3^2L^{-2}L^{1/p'}\|f\|_{p,J} + L^{1/p'}\|f''\|_{p,J} \\ &\leq L^{1/p'}2(3^{2p}L^{-2p}\|f\|_{p,J}^p + \|f''\|_{p,J}^p)^{1/p} \end{aligned} \tag{5}$$

Consequently,

$$\|f'\|_{p,J}^p \leq 18^pL^{-p}\|f\|_{p,J}^p + 2^pL^p\|f''\|_{p,J}^p$$

Since J was an arbitrary I_j, we obtain

$$\begin{aligned} \|f'\|_{p,I}^p &= \sum_j \|f'\|_{p,I_j}^p \leq 18^pL^{-p} \sum_j \|f\|_{p,I_j}^p + 2^pL^p \sum_j \|f''\|_{p,I_j}^p \\ &= 18^pL^{-p}\|f\|_{p,I}^p + 2^pL^p\|f''\|_{p,I}^p \end{aligned} \tag{6}$$

Let $\varepsilon > 0$ be given. Suppose $p = \infty$. If the length of I, written $l(I)$, does not exceed ε, then taking $L = l(I)$ in (3) gives

$$\|f'\|_{\infty,I} \leq \frac{6}{l(I)} \|f\|_{\infty,I} + \varepsilon\|f''\|_{\infty,I} \tag{7}$$

If $\infty > l(I) > \varepsilon$, then I is the union of nonoverlapping intervals I_1 and I_2, where $l(I_1) = k\varepsilon/2$ for some positive integer k and $\varepsilon \geq l(I_2) \geq \varepsilon/2$. Applying (3) to f on the interval I_1 with $L = \varepsilon/2$ yields

$$\|f'\|_{\infty,I_1} \leq \frac{12}{\varepsilon} \|f\|_{\infty,I_1} + \varepsilon\|f''\|_{\infty,I_1} \tag{8}$$

Similarly, taking $L = l(I_2)$ in (3) gives

$$\|f'\|_{\infty,I_2} \leq \frac{12}{\varepsilon} \|f\|_{\infty,I_2} + \varepsilon\|f''\|_{\infty,I_2} \tag{9}$$

Thus, from (8) and (9),

$$\|f'\|_{\infty,I} \leq \frac{12}{\varepsilon} \|f\|_{\infty,I} + \varepsilon\|f''\|_{\infty,I} \tag{10}$$

For $\varepsilon > 0$ fixed, define

$$K(l(I)) = \begin{cases} \dfrac{12}{l(I)} & l(I) \leq \varepsilon \\ \dfrac{12}{\varepsilon} & l(I) > \varepsilon \end{cases}$$

Then K is a nonincreasing function of the length of I, and for $\infty > l(I)$,

$$\|f'\|_{\infty,I} \leq K\|f\|_{\infty,I} + \varepsilon\|f''\|_{\infty,I} \tag{11}$$

by (7) and (10). If $l(I) - \infty$, then (11) still holds upon choosing $L = \varepsilon$ in (3). Hence the lemma is proved for $n = 2$ and $p = \infty$. By making use of (6), a similar argument proves the lemma for $n = 2$ and $1 \leq p < \infty$.

Assume the lemma to be true for $n = j$. Suppose f is in $W^{j+1,p}(I)$, $1 \leq p < \infty$. Then for any compact subinterval J of I, $f^{(j-1)}$ is in $W^{(2,p)}(I)$. Now, the lemma holds for $n = 2$. Hence, given $\eta > 0$, there exists a K_0 depending only on η, p, and the length of J such that $f^{(j)}$ is in $\mathfrak{L}_p(J)$ and

$$\|f^{(j)}\|_{p,J}^p \leq K_0\|f^{(j-1)}\|_{p,J}^p + \eta\|f^{(j+1)}\|_{p,I}^p \tag{12}$$

Given $\eta_1 > 0$, there exists, by the induction hypothesis, a K_1 depending only on η_1, p and the length of J such that

$$\|f^{(i)}\|_{p,J}^p \leq K_1\|f\|_{p,I}^p + \eta_1\|f^{(j)}\|_{p,J}^p \qquad 0 \leq i \leq j-1 \tag{13}$$

Thus, from (12) and (13),

$$\|f^{(j)}\|_{p,J}^p \leq K_0K_1\|f\|_{p,I}^p + K_0\eta_1\|f^{(j)}\|_{p,J}^p + \eta\|f^{(j+1)}\|_{p,I}^p \tag{14}$$

Choosing η_1 so that $1 - K_0\eta_1 \geq \frac{1}{2}$, it follows from (14) that

$$\|f^{(j)}\|_{p,J}^p \leq 2K_0K_1\|f\|_{p,I}^p + 2\eta\|f^{(j+1)}\|_{p,I}^p \tag{15}$$

Since η and η_1 are arbitrarily small and K_0 and K_1 can be chosen to be nonincreasing functions of the length of J, the lemma for $n = j + 1$ follows from (13) and (15). By induction, the lemma is proved. The proof when $p = \infty$ is the same, with $\| \quad \|_\infty$ replacing $\| \quad \|_p^p$.

VI.6.2 Theorem. *Let τ be the differential expression $\tau = \sum_{k=0}^{u} c_k D^k$, where $1/c_n$ and c_k, $0 \leq k \leq n - 1$, are in $\mathfrak{L}_\infty(I)$.*

i. If f is in $\mathfrak{D}(T_{\tau,p,p}) = \{g \mid g \in A_n(I) \cap \mathfrak{L}_p(I),\ \tau g \in \mathfrak{L}_p(I)\}$, $1 \leq p \leq \infty$, then $f^{(k)}$, $0 \leq k \leq n$, is in $\mathfrak{L}_p(I)$.

ii. There exists a K, depending only on p, n, the length of I, and the largest of the numbers $\|1/c_n\|_{\infty,I}$ and $\|c_k\|_{\infty,I}$, $0 \leq k \leq n - 1$, such that for all $f \in \mathfrak{D}(T_{\tau,p,p})$,

$$\|f^{(k)}\|_{p,I}^p \leq K(\|f\|_{p,I}^p + \|\tau f\|_{p,I}^p) \qquad 0 \leq k \leq n,\ 1 \leq p < \infty$$

$$\|f^{(k)}\|_{\infty,I} \leq K(\|f\|_{\infty,I} + \|\tau f\|_{\infty,I}) \qquad 0 \leq k \leq n,\ p = \infty$$

iii. Suppose $1/c_n$ and c_k, $0 \leq k \leq n - 1$, are bounded on the line. Then K, considered as a function of the length of I, can be chosen to be nonincreasing.

Proof. We shall prove the theorem for $1 \leq p < \infty$. The argument for $p = \infty$ is the same, with $\| \quad \|_\infty$ replacing $\| \quad \|_p^p$. Let J be an arbitrary compact subinterval of I. Since

$$f^{(n)} = \frac{\tau f - \sum_{k=0}^{n-1} c_k f^{(k)}}{c_n} \quad \text{a.e.} \tag{1}$$

and $f^{(k)}$, $0 \leq k \leq n - 1$, is continuous on J, it follows from (1) that $f^{(n)}$ is in $\mathfrak{L}_p(J)$ and

$$\begin{aligned}\|f^{(n)}\|_{p,J} &\leq C_I\Big(\|\tau f\|_{p,J} + C_I \sum_{k=0}^{n-1} \|f^{(k)}\|_{p,J}\Big)\\ &\leq C_I\Big[(n+1)\Big(\|\tau f\|_{p,J}^p + C_I{}^p \sum_{k=0}^{n-1} \|f^{(k)}\|_{p,J}^p\Big)^{1/p}\Big]\end{aligned}$$

where C_I is the largest of $\|1/c_n\|_{\infty,I}$ and $\|c_k\|_{\infty,I}$, $0 \le k \le n-1$. Hence

$$\|f^{(n)}\|_{p,J}^p \le K_0 \left(\|\tau f\|_{p,J}^p + \sum_{k=0}^{n-1} \|f^{(k)}\|_{p,J}^p \right) \tag{2}$$

where K_0 depends on C_I, p, and n.

Given $\varepsilon > 0$, there exists, by Lemma VI.6.1, a constant K_1 depending only on ε, p, and the length of J such that

$$\|f^{(k)}\|_{p,J}^p \le K_1 \|f\|_{p,J}^p + \varepsilon \|f^{(n)}\|_{p,J}^p \qquad 0 \le k \le n-1 \tag{3}$$

Thus, from (2) and (3),

$$\begin{aligned} (1 - K_0 n\varepsilon) \|f^{(n)}\|_{p,J}^p &\le K_0 \|\tau f\|_{p,J}^p + K_0 K_1 n \|f\|_{p,J}^p \\ &\le K_0 \|\tau f\|_{p,I}^p + K_0 K_1 n \|f\|_{p,I}^p \end{aligned} \tag{4}$$

In particular, for ε chosen so that $1 - K_0 n \varepsilon \ge \frac{1}{2}$,

$$\|f^{(n)}\|_{p,J}^p \le 2K_0(1 + K_1 n)(\|\tau f\|_{p,I}^p + \|f\|_{p,I}^p) \tag{5}$$

Now I may be expressed as the union of compact subintervals J_1, J_2, . . . , where $J_1 \subset J_2 \subset$ . . . . Furthermore, we know by Lemma VI.6.1 that K_1 may be chosen to be a nonincreasing function of the length of J. Hence, letting $J = J_1$ in (5), it follows that

$$\|f^{(n)}\|_{p,J_i}^p \le 2K_0(1 + K_1 n)(\|\tau f\|_{p,I}^p + \|f\|_{p,I}^p) \qquad i = 1, 2, \ldots$$

Therefore $f^{(n)}$ is in $\mathfrak{L}_p(I)$ and

$$\|f^{(n)}\|_{p,I}^p \le K(\|\tau f\|_{p,I}^p + \|f\|_{p,I}^p) \tag{6}$$

where K depends only on p, n, C_I, and the length of I. Finally, there exists, by Lemma VI.6.1, an M depending only on p and the length of I such that

$$\|f^{(k)}\|_{p,I}^p \le M \|f\|_{p,I}^p + \|f^{(n)}\|_{p,I}^p \qquad 0 \le k \le n-1 \tag{7}$$

Since K_1 and M, considered as functions of the length of I, can be chosen to be nonincreasing, the theorem follows from (6) and (7).

VI.6.3 Corollary. *Let* $\tau = \sum_{k=0}^{n} a_k D^k$, *where each* a_k *is a constant. Then for the differential expression* D, $T_{\tau,p,p} = P(T_{D,p,p})$, *where* P *is the polynomial* $\sum_{k=0}^{n} a_k z^k$ *and* $1 \le p \le \infty$.

Proof. For convenience we write $T_{\tau,p,p}$ as T_τ, and $T_{D,p,p}$ *as* T_D. Suppose $f \in \mathfrak{D}(P(T_D)) = \mathfrak{D}(T_D^n)$. Then $f \in A_n(I) \cap \mathfrak{L}_p(I)$, and $f^{(k)} \in \mathfrak{L}_p(I)$, $0 \leq k \leq n$. Thus $\tau f = \sum_{k=0}^{n} a_k f^{(k)}$ is in $\mathfrak{L}_p(I)$; that is, $f \in \mathfrak{D}(T_\tau)$. Furthermore, $(P(T_D))f = T_\tau f$. Since Theorem VI.6.2 implies $\mathfrak{D}(T_\tau) \subset \mathfrak{D}(T_D^n)$, we may conclude that $T_\tau = P(T_D)$.

VI.6.4 Remarks

i. As seen in Lemma VI.6.1, $W^{n,p}(I)$, $1 \leq p \leq \infty$, concides with $\{g \mid g \in A_n(I),\ g^{(i)} \in \mathfrak{L}_p(I),\ 0 \leq i \leq n\}$. If we define a norm $\|\ \|_{n,p,I}$ on $W^{n,p}(I)$ by

$$\|f\|_{n,p,I} = \sum_{k=0}^{n} \|f^{(k)}\|_{p,I}$$

where $\|\ \|_{p,I}$ is the norm on $\mathfrak{L}_p(I)$, then $W^{n,p}(I)$ is a Banach space. To see this, suppose $\{f_j\}$ is a Cauchy sequence in $W^{n,p}(I)$. Then obviously $\{f_j^{(k)}\}$ is a Cauchy sequence in $\mathfrak{L}_p(I)$ for $0 \leq k \leq n$. Hence $\{f_j^{(k)}\}$ converges in $\mathfrak{L}_p(I)$ to some h_k. By Theorem VI.3.2, $T_{k,p} = T_{D^k,p,p}$ is closed. Since each f_j is in $\mathfrak{D}(T_{k,p})$, $0 \leq k \leq n$, and $T_{k,p}f_j = f_j^{(k)} \to h_k$, it follows that $h_0 = \lim f_j$ is in $\mathfrak{D}(T_{k,p})$ and

$$h_0^{(k)} = T_{k,p}h_0 = \lim_{j\to\infty} T_{k,p}f_j = \lim_{j\to\infty} f_j^{(k)} \qquad 0 \leq k \leq n$$

Thus $\{f_j\}$ converges in $W^{n,p}(I)$ to h_0, proving that $W^{n,p}(I)$ is complete.

ii. If X_{n+1} is the Banach space $\underbrace{\mathfrak{L}_p(I) \times \mathfrak{L}_p(I) \times \cdots \times \mathfrak{L}_p(I)}_{n+1}$ with norm $\|(g_0, g_1, \ldots, g_n)\| = \sum_{i=0}^{n} \|g_i\|_{p,I}$, then $W^{n,p}(I)$ is equivalent to the subspace M_{n+1} of X_{n+1} consisting of those elements $(f, f', \ldots, f^{(n)})$ where f is in $W^{n,p}(I)$. Since $W^{n,p}(I)$ is complete, M_{n+1} is also complete and hence a closed subspace of X_{n+1}. Thus if $1 < p < \infty$, then X_{n+1} is reflexive, implying that $W^{n,p}(I)$ is reflexive. If $1 \leq p < \infty$, then X_{n+1} is separable and therefore $W^{n,p}(I)$ is separable.

VI.7 THE CONSTANT COEFFICIENT AND THE EULER DIFFERENTIAL OPERATORS

VI.7.1 Definition. *Let T be a linear operator with domain and range contained in normed linear space X. The* ***essential spectrum*** *of T,* *written* $\sigma_e(T)$, *is defined by*

$$\sigma_e(T) = \{\lambda \mid \mathcal{R}(\lambda I - T) \text{ is not closed}\}$$

The **essential resolvent** *of* T, *written* $\rho_e(T)$, *is the set of scalars not in* $\sigma_e(T)$.

In the literature one finds other definitions of the essential spectrum.

Balslev and Gamelin [1] have determined $\sigma_e(T_{\tau,p,p})$ and $\kappa(\lambda I - T_{\tau,p,p})$ for $1 < p < \infty$ in Theorems VI.7.2 and VI.7.3.

VI.7.2 Theorem. Let $\tau = \sum_{k=0}^{n} a_k D^k$, *where each* a_k *is a constant, and let* P *be the polynomial* $P(z) = \sum_{k=0}^{n} a_k z^k$. *The maximal operator* $T_{\tau,p}$ *corresponding to* (τ, p, p) *and the interval* $[0, \infty)$ *has the following properties.*

i. $\sigma_e(T_{\tau,p}) = \{P(z) \mid \text{Re } z = 0\} \qquad 1 \leq p \leq \infty$

For $\lambda \in \sigma_e(T_{\tau,p})$, $\mathcal{R}(\lambda I - T_{\tau,p})$ *is a proper subspace dense in* $\mathcal{L}_p([0, \infty))$, $1 \leq p < \infty$.

ii. *For* $\lambda \in \rho_e(T_{\tau,p})$, $\lambda I - T_{\tau,p}$ *is surjective*, $1 \leq p \leq \infty$.

iii. *If* $1 \leq p < \infty$ *and* λ *is in* $\rho_e(T_{\tau,p})$, *then*

$$\kappa(\lambda I - T_{\tau,p}) = \alpha(\lambda I - T_{\tau,p})$$

is the number of zeros of $\lambda - P(z)$, *counted according to their multiplicity, which lie in the half plane* Re $z < 0$.

Proof. Since $\lambda I - T_{\tau,p} = T_{\lambda-\tau,p}$, Theorem VI.2.11 implies $\lambda \in \rho_e(T_{\tau,p})$ if and only if $\lambda I - T_{\tau,p}$ is surjective, $1 \leq p \leq \infty$. Suppose $1 \leq p < \infty$. Now, by Theorems VI.2.7 and IV.2.14 and Corollary VI.6.3,

$$\sigma_e(T_{\tau,p}) = F\sigma(T_{\tau,p}) = F\sigma(P(T_{D,p})) = P(F\sigma(T_{D,p})) \tag{1}$$

Thus the problem reduces to finding $F\sigma(T_{D,p})$. For $f \in \mathcal{L}_p(J)$, $J = [0, \infty)$, the equation $(\lambda I - D)y = f$ has a solution

$$y(t) = -\int_0^t e^{\lambda(t-s)} f(s)\, ds$$

in $A_1(J_0)$ for every compact subinterval J_0 of J.

Suppose Re $\lambda < 0$. Then y is in $\mathcal{L}_p(J)$ by theorem 0.11. Thus $\beta(\lambda I - T_{D,p}) = 0$. Now, $e^{\lambda t}$ is in $\mathfrak{N}(\lambda I - T_{D,p})$, and $\alpha(\lambda I - T_{D,p}) \leq 1$.

Hence $\alpha(\lambda I - T_{D,p}) = 1$ and

$$\kappa(\lambda I - T_{D,p}) = 1 \qquad \text{Re}\,\lambda < 0 \tag{2}$$

Suppose Re $\lambda > 0$. By Theorem VI.2.11, $\lambda I - T_{D,p} = T_{\lambda - D,p}$ is surjective if and only if $T_{\lambda+D,p'}$ is surjective. Given $g \in \mathfrak{L}_{p'}(J)$

$$y(t) = \int_0^t e^{-\lambda(t-s)} g(s)\, ds$$

is a solution in $A_1(J) \cap \mathfrak{L}_{p'}(J)$ to the equation $\lambda y + y' = g$ by Theorem 0.11. Thus $T_{\lambda+D,p'}$ is surjective and consequently $\beta(\lambda I - T_{D,p}) = 0$. Since $e^{\lambda t}$ is a solution to $\lambda y - y' = 0$ on every bounded subinterval of J and $e^{\lambda t}$ is not in $\mathfrak{L}_p(J)$, it follows that $\alpha(\lambda I - T_{D,p}) = 0$. Thus

$$\kappa(\lambda I - T_{D,p}) = 0 \qquad \text{Re}\,\lambda > 0 \tag{3}$$

Suppose Re $\lambda = 0$. Then λ is not in $\rho_e(T_{D,p}) = F\rho(T_{D,p})$; otherwise it would follow from Theorem V.1.6 that $\kappa(\mu I - T_{D,p}) = \kappa(\lambda I - T_{D,p})$ for all μ sufficiently close to λ. But this contradicts the fact that both (2) and (3) hold.

If $p = \infty$, then by Theorem VI.2.7, $\sigma_e(T_{\tau,\infty}) = \sigma_e(T_{\tau^*,1})$, which was just shown to be the set $\{P^*(z) \mid \text{Re}\, z = 0\}$, where P^* is the polynomial $\sum_{k=0}^{n} (-1)^k a_k z^k$ corresponding to $\tau^* = \sum_{k=0}^{n} (-1)^k a_k D^k$. Since $P^*(z) = P(-z)$ for all complex numbers z, it follows that

$$\sigma_e(T_{\tau,\infty}) = \{P(z) \mid \text{Re}\, z = 0\}$$

Since $(\lambda I - T_{\tau,p})' = T_{o,\lambda-\tau^*,p'}$, $1 \le p < \infty$, Theorem II.3.7 implies that $\mathfrak{R}(\lambda I - T_{\tau,p})$ is dense in $\mathfrak{L}_p([0, \infty))$, $1 \le p < \infty$. Thus (*i*) is proved. The rest of the theorem follows from (2), (3), and Theorem IV.2.14.

The classical method of solving the Euler differential equation $\sum_{k=0}^{n} b_k t^k D^k f = g$, where each b_k is a constant, is to introduce a change of variable, thereby transforming the given equation into one of the form $\nu f_1 = g_1$, where ν is a differential expression with constant coefficients. A similar procedure is used in the proof of the next theorem.

VI.7.3 Theorem. *Let* $\tau = \sum_{k=0}^{n} b_k t^k D^k$, *where each* b_k *is a constant, and let* P *be the polynomial*

$$P(z) = b_0 + \sum_{k=1}^{n} b_k \prod_{j=0}^{k-1} \left[z - \left(j + \frac{1}{p} \right) \right]$$

with $1/p = 0$ when $p = \infty$. The maximal operator $T_{\tau,p}$ corresponding to (τ, p, p) and the interval $[1, \infty)$ has the following properties.

i. $\sigma_e(T_{\tau,p}) = \{P(z) \mid \text{Re } z = 0\} \qquad 1 \leq p \leq \infty$

For $\lambda \in \sigma_e(T_{\tau,p})$, $\mathcal{R}(\lambda I - T_{\tau,p})$ is a proper dense subspace of $\mathcal{L}_p([1, \infty))$.

ii. For $\lambda \in \rho_e(T_{\tau,p})$, $\lambda I - T_{\tau,p}$ is surjective, $1 \leq p \leq \infty$.

iii. If $1 \leq p < \infty$ and λ is in $\rho_e(T_{\tau,p})$, then

$$\kappa(\lambda I - T_{\tau,p}) = \alpha(\lambda I - T_{\tau,p})$$

is the number of zeros of $\lambda - P(z)$, counted according to their multiplicity, which lie in the half plane Re $z < 0$.

Proof. Let $J_0 = [0, \infty)$ and $J_1 = [1, \infty)$. Then $\mathcal{L}_p(J_1)$ is equivalent to $\mathcal{L}_p(J_0)$ under the map $\eta : \mathcal{L}_p(J_1) \to \mathcal{L}_p(J_0)$ defined by

$$(\eta f)(s) = e^{s/p} f(e^s) \qquad 0 \leq s < \infty \tag{1}$$

with the understanding that $e^{s/p} = 1$ when $p = \infty$. The statement follows from the well-known theorem concerning integration by substitution (cf. McShane [1], 38.4, page 214). The inverse map η^{-1} is given by

$$f(t) = (\eta^{-1} g)(t) = t^{-1/p} g(\log t) \qquad 1 \leq t < \infty \tag{2}$$

Since e^s and $\log t$ are monotone increasing and infinitely differentiable on the line and J_0, respectively, it follows from (1) and (2) that

$$\eta(A_n(J_1) \cap \mathcal{L}_p(J_1)) = A_n(J_0) \cap \mathcal{L}_p(J_0) \tag{3}$$

(cf. McShane [1], 9.3, page 51). Suppose that f is in $A_n(J_1)$. Then $\eta f = g$ is in $A_n(J_0)$, and by (2),

$$\begin{aligned} f'(t) &= t^{-1-1/p} g'(s)]_{s=\log t} - \frac{1}{p} t^{-1-1/p} g(s)]_{s=\log t} \\ &= t^{-1-1/p} \left[\left(\frac{d}{ds} - \frac{1}{p} \right) g(s) \right]_{s=\log t} \end{aligned}$$

It follows by induction that for $1 \leq k \leq n$,

$$f^{(k)}(t) = t^{-k-1/p} \left[\left(\frac{d}{ds} - \left(\frac{1}{p} + k - 1 \right) \right) \left(\frac{d}{ds} - \left(\frac{1}{p} + k - 2 \right) \right) \cdots \left(\frac{d}{ds} - \frac{1}{p} \right) g(s) \right]_{s=\log t}$$

Hence

$$(4) \quad (\tau f)(t) = t^{-1/p}\left\{b_0 + \sum_{k=1}^{n} b_k \prod_{j=0}^{k-1}\left[\left(\frac{d}{ds} - \left(\frac{1}{p} + j\right)\right)g(s)\right]_{s=\log t}\right\}$$

Let ν be the differential expression

$$\nu = P\left(\frac{d}{ds}\right) = b_0 + \sum_{k=1}^{n} b_k \prod_{j=0}^{k-1}\left(\frac{d}{ds} - \left(\frac{1}{p} + j\right)\right)$$

and let $T_{\nu,p,p}$ be the maximal operator corresponding to (ν, p, p) with respect to the interval J_0. Then from (4) and the definition of η

$$(5) \qquad \eta(\tau f) = \nu g = \nu\eta f$$

Thus, by (3) and (5), $T_{\tau,p,p} = \eta^{-1}T_{\nu,p,p}\,\eta$, or equivalently, for any scalar λ,

$$\lambda I - T_{\tau,p,p} = \eta^{-1}(\lambda I - T_{\nu,p,p})\eta$$

Since η is a linear isometry, $T_{\tau,p,p}$ and $T_{\nu,p,p}$ have the same essential spectrum and $\kappa(\lambda I - T_{\tau,p,p}) = \kappa(\lambda I - T_{\nu,p,p})$ for $\lambda \in \rho_e(T_{\tau,p,p})$. The theorem now follows from Theorem VI.7.2.

The results in the remaining sections of this chapter generalize those due to Balslev and Gamelin [1] for the constant coefficient and the Euler differential operators.

VI.8 PERTURBATIONS OF THE BOUNDED COEFFICIENT AND THE EULER OPERATORS

VI.8.1 Theorem. *Let J be the interval $[a, \infty)$ and let T be the maximal operator corresponding to (τ, p, p), $1 < p < \infty$, where*

$$\tau = \sum_{k=0}^{n} a_k D^k \qquad \frac{1}{a_n} \in \mathfrak{L}_\infty(J)$$

$$a_k \in \mathfrak{L}_\infty(J)$$

$$0 \le k \le n$$

Let B be the maximal operator corresponding to (ν, p, p), $1 < p < \infty$, where

$$\nu = \sum_{k=0}^{n-1} b_k D^k \qquad \textit{each } b_k \textit{ measurable on } J$$

Then

i. *B is T-bounded if and only if each b_k is locally in $\mathcal{L}_p(J)$ and*

$$\sup_{a \leq s < \infty} \int_s^{s+1} |b_k(t)|^p \, dt < \infty \qquad 0 \leq k \leq n-1$$

In this case, given $\varepsilon > 0$, there exists a K, depending only on ε and p, such that for all $f \in \mathfrak{D}(T)$,

$$\|Bf\|_{p,J} \leq K\|f\|_{p,J} + \varepsilon\|Tf\|_{p,J}$$

Furthermore, $T_{\tau+\nu} = T + B$, where $T_{\tau+\nu}$ is the maximal operator corresponding to $(\tau + \nu, p, p)$.

ii. *B is T-compact if and only if each b_k is locally in $\mathcal{L}_p(J)$ and*

$$\lim_{s \to \infty} \int_s^{s+1} |b_k(t)|^p \, dt = 0 \qquad 0 \leq k \leq n-1$$

In this case, T and $T_{\tau+\nu}$ have the same essential spectrum. For $\lambda \in \rho_e(T)$, $\kappa(\lambda I - T) = \kappa(\lambda I - T_{\tau+\nu})$.

Before proving the theorem, we need the following lemma.

VI.8.2 Lemma. *Given interval I with length greater than 1 and $\epsilon > 0$, there exists a K, depending only on ε, the length of I, and p, $1 < p < \infty$, such that for all b locally in $\mathcal{L}_p(I)$ and all f in the domain of the maximal operator T corresponding to (D, p, p),*

$$\|bf\|_{p,I}^p \leq (\varepsilon\|f'\|_{p,I}^p + K\|f\|_{p,I}^p) \sup_{[s,s+1] \subset I} \int_s^{s+1} |b(t)|^p \, dt$$

K may be chosen to be nonincreasing as a function of the length of I.

Proof. Let I_1 and I_2 be nonoverlapping subintervals of I such that $I = I_1 \cup I_2$, with I_1 "to the left" of I_2. For $\eta > 0$ such that $t + \eta$ and $t - \eta$ are in I for all $t \in I_1$ and I_2, respectively, choose $\varphi \in C'([0, \eta])$ so that $0 \leq \varphi(t) \leq 1$ on $[0, \eta]$, $\varphi(0) = 1$, and $\varphi(\eta) = 0$. For $f \in \mathfrak{D}(T)$ and $t \in I_1$,

$$f(t) = -\int_0^\eta \frac{d}{ds}(\varphi(s)f(t+s)) \, ds = -\int_0^\eta \varphi(s)f'(t+s) \, ds - \int_0^\eta \varphi'(s)f(t+s) \, ds$$

Letting $M = \max_{0 \le s \le \eta} |\varphi'(s)|$, we obtain from Hölder's inequality

$$\begin{aligned}(1)\quad |f(t)| &\le \int_0^\eta |f'(t+s)|\,ds + M \int_0^\eta |f(t+s)|\,ds \\ &\le \eta^{1/p'} \left[\left(\int_0^\eta |f'(t+s)|^p\,ds \right)^{1/p} + M \left(\int_0^\eta |f(t+s)|^p\,ds \right)^{1/p} \right] \\ &\le 2\eta^{1/p'} \left(\int_0^\eta |f'(t+s)|^p\,ds + M^p \int_0^\eta |f(t+s)|^p\,ds \right)^{1/p}\end{aligned}$$

Taking η sufficiently small, it follows that there exists a K_1 depending only on ε, p, and the length of I such that

$$\begin{aligned}(2)\qquad |f(t)|^p &\le \frac{\varepsilon}{2} \int_0^\eta |f'(t+s)|^p\,ds + K_1 \int_0^\eta |f(t+s)|^p\,ds \\ &= \frac{\varepsilon}{2} \int_t^{t+\eta} |f'(s)|^p\,ds + K_1 \int_t^{t+\eta} |f(s)|^p\,ds \qquad t \in I_1\end{aligned}$$

From the conditions put on η we see that as the length of I increases, η can be kept fixed so that (1) still holds. Thus K_1 can be chosen to be nonincreasing as a function of the length of I. Letting a be the left endpoint of I_1, Fubini's theorem and (2) imply that for $t \in I_1$ and η sufficiently small,

$$\begin{aligned}(3)\quad \int_{I_1} |b(t)f(t)|^p\,dt &\le \int_{I_1} \int_t^{t+\eta} |b(t)|^p \left(\frac{\varepsilon}{2} |f'(s)|^p + K_1 |f(s)|^p \right) ds\,dt \\ &\le \int_I \frac{\varepsilon}{2} |f'(s)|^p + K_1 |f(s)|^p\,ds \int_{\max(s-\eta,a)}^s |b(t)|^p\,dt \\ &\le \left(\frac{\varepsilon}{2} \|f'\|_{p,I}^p + K_1 \|f\|_{p,I}^p \right) \sup_{[s,s+1]\subset I} \int_s^{s+1} |b(t)|^p\,dt\end{aligned}$$

For $t \in I_2$,

$$f(t) = -\int_0^\eta \frac{d}{ds} (\varphi(s) f(t-s))\,ds$$

Thus, by the argument used to establish (2),

$$|f(t)|^p \le \frac{\varepsilon}{2} \int_{t-\eta}^t |f'(s)|^p\,ds + K_1 \int_{t-\eta}^t |f(s)|^p\,ds \qquad t \in I_2$$

As in (3),

$$\begin{aligned}(4)\quad \int_{I_2} |b(t)f(t)|^p\,dt &\le \int_I \frac{\varepsilon}{2} |f'(s)|^p + K_1 |f(s)|^p\,ds \int_s^{s+\eta} |b(t)|^p\,dt \\ &\le \left(\frac{\varepsilon}{2} \|f'\|_{p,I}^p + K_1 \|f\|_{p,I}^p \right) \sup_{[s,s+1]\subset I} \int_s^{s+1} |b(t)|^p\,dt\end{aligned}$$

The lemma now follows from (3) and (4).

Proof of the theorem. Suppose b_k is locally in $\mathfrak{L}_p(J)$ and

$$\sup_{a \le s < \infty} \int_s^{s+1} |b_k(t)|^p \, dt < \infty \qquad 0 \le k \le n-1$$

Let J_1 be any subinterval of J of length greater than 1. Since any non-negative numbers c and d satisfy $(c^p + d^p)^{1/p} \le c + d$, it follows from Lemmas VI.8.2 and VI.6.1 and Theorem VI.6.2 that given $\mu > 0$ and $\eta > 0$, there exist constants K_0 and K_1, depending only on p and μ and p and η, respectively, such that for all $f \in \mathfrak{D}(T)$,

$$\|b_k f^{(k)}\|_{p,J_1} \le \mu \|f^{(k+1)}\|_{p,J_1} + K_0 \|f^{(k)}\|_{p,J_1} \qquad 0 \le k \le n-1 \tag{1}$$

$$\|f^{(k)}\|_{p,J_1} \le \eta \|f^{(n)}\|_{p,J_1} + K_1 \|f\|_{p,J_1} \qquad 0 \le k \le n-1 \tag{2}$$

Using equations (1) and (2), a simple computation shows the existence of a K_2, depending only on p and ε, such that for all $f \in \mathfrak{D}(T)$,

$$\|b_k f^{(k)}\|_{p,J_1} \le \varepsilon \|f^{(n)}\|_{p,J_1} + K_2 \|f\|_{p,J_1} \qquad 0 \le k \le n-1 \tag{3}$$

It follows from Theorem VI.6.2 that there exists a constant C, depending only on p, such that for all $f \in \mathfrak{D}(T)$,

$$\|f^{(n)}\|_{p,J_1} \le C(\|f\|_{p,J_1} + \|\tau f\|_{p,J_1}) \tag{4}$$

We may derive from (3) and (4) an inequality of the form

$$\|\nu f\|_{p,J_1} \le \sum_{k=0}^{n-1} \|b_k f^{(k)}\|_{p,J_1} \le K \|f\|_{p,J_1} + \varepsilon \|\tau f\|_{p,J_1} \qquad f \in \mathfrak{D}(T) \tag{5}$$

where K depends only on p and ε.

It is clear from (5) that $\mathfrak{D}(T)$ is contained in $\mathfrak{D}(B)$ and that B is T-bounded. To prove $T_{\tau+\nu} = T + B$, it suffices to show that $\mathfrak{D}(T_{\tau+\nu}) = \mathfrak{D}(T)$. If $f \in \mathfrak{D}(T)$, τf and νf are in $\mathfrak{L}_p(J)$ by (5), which implies $f \in \mathfrak{D}(T_{\tau+\nu})$. On the other hand, if f is in $\mathfrak{D}(T_{\tau+\nu})$, then $(\tau + \nu)f$ is in $\mathfrak{L}_p(J)$. Now, $f^{(j)}$, $0 \le j \le n-1$, is continuous on each compact subinterval J_1 of J. Hence νf and $\tau f = (\tau + \nu)f - \nu f$ are in $\mathfrak{L}_p(J_1)$. By considering only those J_1 of length greater than 1 we have from (5) and the triangular inequality that

$$\|\nu f\|_{p,J_1} \le K \|f\|_{p,J_1} + \varepsilon(\|(\tau + \nu)f\|_{p,J_1} + \|\nu f\|_{p,J_1})$$

whence

$$\|\nu f\|_{p,J_1} \le \frac{K}{1-\varepsilon} \|f\|_{p,J} + \frac{\varepsilon}{1-\varepsilon} \|(\tau + \nu)f\|_{p,J} \qquad 0 < \varepsilon < 1 \tag{6}$$

Since (6) holds for all subintervals J_1 of length greater than 1, it follows that νf is in $\mathfrak{L}_p(J)$. Consequently, $\tau f = (\tau + \nu)f - \nu f$ is in $\mathfrak{L}_p(J)$, implying that f is in $\mathfrak{D}(T)$. Thus $T + B = T_{\tau+\nu}$ and (5) implies

$$\|Bf\|_{p,J} \leq K\|f\|_{p,J} + \varepsilon\|Tf\|_{p,J} \qquad f \in \mathfrak{D}(T)$$

where K depends only on p and ε.

Suppose B is T-bounded. Let φ be a function in $C_0^\infty(E^1)$ such that $\varphi = 1$ on $[0, 1]$. For each $s \geq 0$, define φ_s to be the function φ translated so as to be 1 on $[s, s + 1]$; that is, $\varphi_s(t) = \varphi(t - s)$. Now

$$B\varphi_s = \sum_{k=0}^{n-1} b_k\varphi_s^{(k)} = b_0 \qquad \text{on } [s, s + 1]$$

Hence

$$\text{(7)} \quad \left(\int_s^{s+1} |b_0|^p\right)^{1/p} \leq \|B\varphi_s\|_{p,J} \leq \|B\|_T\|\varphi_s\|_T = \|B\|_T(\|\varphi_s\|_{p,J} + \|T\varphi_s\|_{p,J})$$

where $\| \quad \|_T$ is the T-norm on $\mathfrak{D}(T)$. Since

$$\int_{-\infty}^{\infty} \frac{d^k}{dt^k}\, \varphi_s(t)\, dt = \int_{-\infty}^{\infty} \frac{d^k}{dt^k}\, \varphi(t - s)\, dt = \int_{-\infty}^{\infty} \varphi^{(k)}(t)\, dt$$

it follows from (7) that

$$\left(\int_s^{s+1} |b_0(t)|^p\, dt\right)^{1/p} \leq \|B\|_T \left[\left(\int_{-\infty}^{\infty} |\varphi(t)|^p\, dt\right)^{1/p} + \sum_{k=0}^{n} \|a_k\|_\infty \left(\int_{-\infty}^{\infty} |\varphi^{(k)}(t)|^p\, dt\right)^{1/p}\right]$$

Thus

$$\sup_{a \leq s < \infty} \int_s^{s+1} |b_0(t)|^p\, dt < \infty$$

Suppose

$$\text{(8)} \qquad \sup_{a \leq s < \infty} \int_s^{s+1} |b_i(t)|^p\, dt < \infty \qquad 0 \leq i \leq k - 1 < n - 1$$

We show that (8) holds for $0 \leq i \leq k$. Define h to be in $C_0^\infty(E^1)$ so that $h^{(k)}$ is 1 on $[0, 1]$. This can be done, for example, by defining h to be φg, where $g = t^k/k!$. Let $h_s(t) = h(t - s)$. Then $h_s^{(k)}(t) = 1$ on $[s, s + 1]$, and

$$Bh_s = b_0h_s + b_1h_s^{(1)} + \cdots + b_{k-1}h_s^{(k-1)} + b_k \qquad \text{on } [s, s + 1]$$

Hence

$$(9)\quad \left(\int_s^{s+1} |b_k(t)|^p\,dt\right)^{1/p} \leq \|B\|_T\|h_s\|_T + M \sum_{i=0}^{k-1} \left(\int_s^{s+1} |b_i(t)|^p\,dt\right)^{1/p}$$

where $M \geq \sup_{-\infty<t<\infty} |h^{(i)}(t)|,\ 0 \leq i \leq n-1$. Since

$$\|h_s\|_T \leq \left(\int_{-\infty}^{\infty} |h(t)|^p\,dt\right)^{1/p} + \sum_{i=0}^{n} \|a_i\|_\infty \left(\int_{-\infty}^{\infty} |h^{(i)}(t)|^p\,dt\right)^{1/p}$$

equation (9), together with (8), implies

$$\sup_{a\leq s<\infty} \int_s^{s+1} |b_i(t)|^p\,dt < \infty \qquad 0 \leq i \leq k$$

Thus (8) holds for $0 \leq i \leq n$ by induction. Hence (i) is proved.

Proof of (ii). Suppose b_k is locally in $\mathfrak{L}_p(J)$ and

$$(10)\qquad \lim_{s\to\infty} \int_s^{s+1} |b_k(t)|^p\,dt = 0 \qquad 0 \leq k \leq n-1$$

For each positive integer $N > a$ define B_N on $\mathfrak{D}(T)$ by

$$B_N f = \begin{cases} Bf & \text{on } [a, N] \\ 0 & \text{on } (N, \infty) \end{cases}$$

We show that B_N converges to B in the space of bounded operators on $\mathfrak{D}(T)$ with the T-norm. It follows from Lemma VI.8.2 and Theorem VI.6.2 that there exist constants K and C depending only on p such that for all $f \in \mathfrak{D}(T)$,

$$\begin{aligned}(11)\quad \|Bf - B_N f\| &\leq \sum_{k=0}^{n-1} \left(\int_N^{\infty} |b_k(t) f^{(k)}(t)|^p\right)^{1/p} \\ &\leq \sum_{k=0}^{n-1} (\|f^{(k+1)}\|_{p,J} \\ &\qquad + K\|f^{(k)}\|_{p,J}) \left(\sup_{N\leq s<\infty} \int_s^{s+1} |b_k(t)|^p\,dt\right)^{1/p}\end{aligned}$$

$$(12)\qquad \|f^{(k)}\|_{p,J} \leq C(\|f\|_{p,J} + \|Tf\|_{p,J}) \qquad 0 \leq k \leq n$$

Considering $\mathfrak{D}(T)$ to be the Banach space with the T-norm, it is clear from (10), (11), and (12) that $B_N \to B$ in $[\mathfrak{D}(T), \mathfrak{L}_p(J)]$. To prove that B is T-compact, it suffices to show that each B_N is T-compact. Suppose $\{f_j\}$

is T-bounded. Then $\{f_j^{(k)}\}$, $0 \leq k \leq n-1$, is uniformly bounded on J, since equation (2) in the proof of lemma VI.8.2 implies the existence of a constant K depending only on p such that

$$|f_j^{(k)}(t)| \leq \|f_j^{(k+1)}\|_{p,J} + K\|f_j^{(k)}\|_{p,J}\cdot \qquad 0 \leq k \leq n-1$$

which, together with (12), shows that $\{f_j^{(k)}\}$ is uniformly bounded on J. Furthermore, $\{f_j^{(k)}\}$ is equicontinuous on [a, N], $0 \leq k \leq n-1$, since, again by (12),

$$\begin{aligned}|f_j^{(k)}(t) - f_j^{(k)}(s)| = \left|\int_s^t f_j^{(k+1)}(\xi)\, d\xi\right| &\leq |t-s|^{1/p'}\|f_j^{(k+1)}\|_{p,J} \\ &\leq C|t-s|^{1/p'}(\|f_j\|_{p,J} + \|Tf_j\|_{p,J})\end{aligned}$$

Hence, by the Ascoli-Arzelá theorem, $\{f_j\}$ has a subsequence $\{f_{j,0}\}$ which converges uniformly on $[a, N]$, and $\{f'_{j,0}\}$ has a subsequence $\{f'_{j,1}\}$ which converges uniformly on $[a, N]$. Thus $\{f_{j,1}\}$ and $\{f'_{j,1}\}$ converge uniformly on $[a, N]$. Continuing in this manner, a subsequence $\{g_j\}$ of $\{f_j\}$ is obtained such that for $0 \leq k \leq n-1$, $\{g_j^{(k)}\}$ converges uniformly on $[a, N]$. Thus

$$\begin{aligned}\|B_N g_i - B_N g_j\|_{p,J} &\leq \sum_{k=0}^{n-1}\left(\int_a^N |b_k(t)|^p |g_i^{(k)}(t) - g_j^{(k)}(t)|^p\right)^{1/p} \\ &\leq \sum_{k=0}^{n-1}\left(\sup_{a\leq t\leq N} |g_i^{(k)}(t) - g_j^{(k)}(t)|\right)\left(\int_a^N |b_k(t)|^p\, dt\right)^{1/p}\end{aligned}$$

implying that $\{B_N g_j\}$ converges in $\mathfrak{L}_p(J)$ as $j \to \infty$. Hence B_N is T-compact and therefore B is T-compact. Clearly, any set which is $\lambda I - T$-bounded is T-bounded. Moreover, it follows from (i) that any set which is $\lambda I - T - B$-bounded is also T-bounded. Hence B is both $\lambda I - T$- and $\lambda I - T - B$-compact. Thus it follows from Theorem V.3.7 that $\rho_e(T) = \rho_e(T + B) = \rho_e(T_{\tau+\nu})$ and $\lambda \in \rho_e(T)$ implies

$$\kappa(\lambda I - T) = \kappa(\lambda I - T_{\tau+\nu})$$

Conversely, suppose B is T-compact. Assume there exists some $\varepsilon > 0$ and a sequence $\{s_k\}$ of positive numbers converging to ∞ such that

$$\int_{s_k}^{s_k+1} |b_0(t)|^p\, dt \geq \varepsilon \qquad 1 \leq k < \infty$$

Let $\{\varphi_s\}$ be the functions defined in the proof of (i). Then

$$\varepsilon^{1/p} \leq \left(\int_{s_k}^{s_k+1} |b_0(t)|^p\, dt\right)^{1/p} \leq \|B\varphi_{s_k}\|_{p,J} \tag{13}$$

Now

$$\|\varphi_{s_k}\|_{p,J} + \|T\varphi_{s_k}\|_{p,J} \le \left(\int_{-\infty}^{\infty} |\varphi(t)|^p\,dt\right)^{1/p} + \sum_{k=0}^{n} \|a_k\|_{\infty,J} \left(\int_{-\infty}^{\infty} |\varphi^{(k)}(t)|^p\,dt\right)^{1/p}$$

which means that $\{\varphi_{s_k}\}$ is T-bounded. Thus, by passing to a subsequence if necessary, we may assume that $\{B\varphi_{s_k}\}$ converges in $\mathfrak{L}_p(J)$ to some y. Given a finite interval $J_0 \subset J$, $\varphi_{s_k} = 0$ on J_0 for all s_k sufficiently large, since φ has compact support. For such s_k,

$$\int_{J_0} |y(t)|^p = \int_{J_0} |y(t) - B\varphi_{s_k}|^p \to 0 \qquad \text{as } s_k \to \infty$$

Since J_0 was an arbitrary finite subinterval of I, y must be the zero element in $\mathfrak{L}_p(J)$; that is, $B\varphi_{s_k} \to 0$. This, however, contradicts (13). Hence $\lim_{s\to\infty} \int_s^{s+1} |b_0(t)|^p = 0$. The proof that

$$\lim_{s\to\infty} \int_s^{s+1} |b_k(t)|^p = 0 \qquad 0 \le k \le n-1$$

proceeds by induction as in the proof of (*i*). This concludes the proof of the theorem.

For results related to Theorem VI.8.1, the reader is referred to Birman [1].

VI.8.3 Corollary. *Given $J = [a, \infty)$ and $1 < p < \infty$, let T and L be the maximal operators corresponding to (τ, p, p) and (l, p, p), respectively, where*

$$\tau = \sum_{k=0}^{n} a_k D^k \qquad \frac{1}{a_n} \text{ and } a_k \text{ in } \mathfrak{L}_\infty(J),\ 0 \le k \le n$$

$$l = \tau + \sum_{k=0}^{n} b_k D^k \qquad b_n \text{ and } \frac{1}{(a_n + b_n)} \text{ in } \mathfrak{L}_\infty(J)$$

$$b_k \text{ locally in } \mathfrak{L}_p(J)$$

$$\lim_{s\to\infty} \int_s^{s+1} |b_k|^p = 0,\ 0 \le k \le n$$

Then $\mathfrak{D}(T) = \mathfrak{D}(L)$, $\sigma_e(T) = \sigma_e(L)$, and $\lambda \in \rho_e(T)$ implies

$$\kappa(\lambda I - T) = \kappa(\lambda I - L)$$

Proof. In view of Theorem VI.8.1, it suffices to prove the corollary for $l = \tau + b_n D^n$. Theorem VI.6.2 implies that $\mathfrak{D}(T) = \mathfrak{D}(L)$. For any scalar λ,

$$(1) \qquad \lambda - l = \lambda - \tau + \frac{b_n}{a_n}\left(\lambda - \tau + \sum_{k=0}^{n-1} a_k D^k - \lambda\right)$$

$$= \left(1 + \frac{b_n}{a_n}\right)(\lambda - \tau) + \sum_{k=0}^{n-1} b_n \frac{a_k}{a_n} D^k - b_n \frac{\lambda}{a_n}$$

Let A_λ be the maximal operator corresponding to $(1 + b_n/a_n)(\lambda - \tau)$. It follows from (1) and Theorem VI.8.1 that $\mathfrak{R}(\lambda I - L)$ is closed if and only if $\mathfrak{R}(A_\lambda)$ is closed, in which case $\kappa(A_\lambda) = \kappa(\lambda I - L)$. Since

$$\alpha(A_\lambda) = \alpha(\lambda I - T) \qquad \text{and} \qquad \beta(A_\lambda) = \beta(\lambda I - T)$$

and since $\mathfrak{R}(A_\lambda)$ is closed if and only if $\mathfrak{R}(\lambda I - T)$ is closed, the theorem follows.

***VI.8.4* Corollary.** *Let $J = [1, \infty)$. For $1 < p < \infty$, let T and L be the maximal operators corresponding to (τ, p, p) and (l, p, p), respectively, where*

$$(\tau f)(t) = \sum_{k=0}^{n} a_k t^k f^{(k)}(t) \qquad \textit{each } a_k \textit{ constant, } a_n \neq 0$$

$$(lf)(t) = (\tau f)(t) + \sum_{k=0}^{n} b_k(t) t^k f^{(k)}(t)$$

the coefficients b_k being subject to the following conditions:

i. b_n and $1/(a_n + b_n)$ are in $\mathfrak{L}_\infty(J)$.

ii. b_k is locally in $\mathfrak{L}_p(J)$ with $\lim_{s\to\infty} \int_{e^s}^{e^{s+1}} \frac{1}{t} |b_k(t)|^p = 0 \qquad 0 \leq k \leq n.$

Then $\mathfrak{D}(T) = \mathfrak{D}(L)$, $\sigma_e(T) = \sigma_e(L)$, and for $\lambda \in \rho_e(T)$,

$$\kappa(\lambda I - T) = \kappa(\lambda I - L)$$

Proof. As in the proof of Theorem VI.7.3, we let η be the linear isometry from $\mathfrak{L}_p([1, \infty))$ onto $\mathfrak{L}_p([0, \infty))$ defined by $(\eta f)(s) = e^{s/p} f(e^s)$, $0 \leq s < \infty$. Then, as was shown, for $f \in \mathfrak{D}(T)$ and $g = \eta f$,

$$(1) \quad t^k f^{(k)}(t) = t^{-1/p} \left\{ \prod_{j=0}^{k-1} \left(\frac{d}{ds} - \left(\frac{1}{p} + j\right)\right) g(s) \right]_{s=\log t} \right\} \qquad 1 \leq k \leq n$$

and

$$(\tau f)(t) = t^{-1/p} \left\{ a_0 + \sum_{k=1}^{n} a_k \prod_{j=0}^{k-1} \left(\frac{d}{ds} - \left(\frac{1}{p} + j \right) \right) g(s) \right]_{s=\log t} \right\}$$

Furthermore, (1) implies that

$$\sum_{k=0}^{n} b_k(t) t^k f^{(k)}(t) = t^{-1/p} \left\{ b_0(e^s) + \sum_{k=1}^{n} b_k(e^s) \prod_{j=0}^{k-1} \left(\frac{d}{ds} - \left(\frac{1}{p} + j \right) \right) g(s) \right]_{s=\log t} \right\}$$

As in the proof of Theorem VI.7.3, it follows that

$$T = \eta^{-1} M \eta \tag{2}$$

and

$$L = \eta^{-1} V \eta \tag{3}$$

where M and V are the maximal operators with domains contained in $\mathfrak{L}_p([0, \infty))$ corresponding to the differential expressions

$$\mu = a_0 + \sum_{k=1}^{n} a_k \prod_{j=0}^{k-1} \left[\frac{d}{ds} - \left(\frac{1}{p} + j \right) \right] \tag{4}$$

and

$$\nu = \mu + b_0(e^s) + \sum_{k=1}^{n} b_k(e^s) \prod_{j=0}^{k-1} \left[\frac{d}{ds} - \left(\frac{1}{p} + j \right) \right] \tag{5}$$

respectively. Noting that for $x \geq 0$,

$$\int_x^{x+1} |b_k(e^s)|^p \, ds = \int_{e^x}^{e^{x+1}} \frac{1}{t} |b_k(t)|^p \, dt \qquad 0 \leq k \leq n$$

the theorem follows from (2) to (5) and Corollary VI.8.3.

As an application of the above theorem, the following result, due to Rota [2], is obtained.

VI.8.5 Example. For $J = [1, \infty)$ and $1 < p < \infty$, let L be the maximal operator corresponding to (l, p, p), where l is the Riemann differential expression

$$(1)\qquad (lf)(t) = t(t+1)f^{(2)}(t) + (at+b)f'(t) + \frac{ct^2+dt+e}{t(t+1)}f(t)$$

with a, b, c, d and e constants. We write (1) in the form

$$(2)\qquad (lf)(t) = t^2f^{(2)}(t) + atf'(t) + cf(t) + (\nu f)(t)$$

where

$$(3)\qquad (\nu f)(t) = b_2(t)t^2f^{(2)}(t) + b_1(t)tf'(t) + b_0(t)f(t)$$

with $b_2(t) = 1/t$, $b_1(t) = b/t$ and $b_0(t) = (d-c)/(t+1) + e/t(t+1)$. Clearly, b_0, b_1 and b_2 satisfy the conditions of Corollary VI.8.4. Hence $\sigma_e(L) = \sigma_e(T)$, where T is the maximal operator corresponding to $t^2f^{(2)}(t) + atf'(t) + cf(t) = (\tau f)(t)$.

Furthermore, if λ is in $\rho_e(T)$, then $\kappa(\lambda I - T) = \kappa(\lambda I - L)$. Thus, by Theorem VI.7.3,

$$(4)\quad \sigma_e(L) = \left\{-r^2 + \frac{1}{p^2} + \frac{(1-a)}{p} + c + ir\left(a - 1 - \frac{2}{p}\right) \quad -\infty < r < \infty\right\}$$

For $\lambda \in \rho_e(L)$, $\kappa(\lambda I - L)$ is the number of zeros of the polynomial

$$(5)\qquad \lambda - c - a\left(z - \frac{1}{p}\right) - \left(z - \frac{1}{p}\right)\left(z - 1 - \frac{1}{p}\right)$$

counted according to their multiplicity, which lie in the halfplane $\operatorname{Re} z < 0$. In particular, if $a = c = 0$, then (4) and (5) imply that 0 is in $\rho_e(L)$ and $\kappa(L) = 0$, since the zeros of the polynomial in (5) with $\lambda = 0$ are $1/p$ and $1 + 1/p$.

VI.8.6 Theorem. *Let J be compact and let T be the maximal operator corresponding to (τ, p, p), $1 < p < \infty$, where*

$$\tau = \sum_{k=0}^{n} b_kD^k \qquad b_k \in \mathfrak{L}_p(J)$$

$$0 \le k \le n-1$$

$$b_n \text{ and } \frac{1}{b_n} \text{ in } \mathfrak{L}_\infty(J)$$

Suppose $B: \mathfrak{L}_p(J) \to \mathfrak{L}_p(J)$ *is a bounded linear operator with* $\mathfrak{D}(B) \supset \mathfrak{D}(A)$, *where* A *is the maximal operator corresponding to* (D^n, p, p) *and* J. *Then* $T + B$ *is a Fredholm operator with index* n *whenever*

$$\|B\| < \frac{(n-1)![(n-1)p' + 1]^{1/p'}(np)^{1/p}\|b_n\|_\infty}{l^n}$$

where l *is the length of* J *and* p' *is conjugate to* p.

Proof. Let K be the maximal operator corresponding to (ν, p, p), where $\nu = \sum_{k=0}^{n-1} (b_k/b_n)D^k$. An argument similar to that given in the proof of Theorem VI.8.1 shows that K is A-compact and that $T = b_n(A + K)$. Since B is bounded, it follows that K is $A + (1/b_n)B$-compact. Hence

$$\kappa(T + B) = \kappa\left(A + K + \frac{1}{b_n} B\right) = \kappa\left(A + \frac{1}{b_n} B\right) \tag{1}$$

whenever the indices exist. Now, for $\left\|\frac{1}{b_n} B\right\| < \gamma(A)$,

$$\kappa\left(A + \frac{1}{b_n} B\right) = \kappa(A) = n \tag{2}$$

The theorem now follows from (1), (2) and Example IV.1.4.

As another application of perturbation theory, the following result, stated somewhat classically, is obtained.

VI.8.7 Theorem. *Given* J *a compact interval, let* k *be a bounded measurable function on* $J \times J$, $a_j \in \mathfrak{L}_1(J)$, $0 \leq j \leq n - 1$, *and* $1/a_n \in \mathfrak{L}_\infty(J)$. *Then, except perhaps for isolated scalars* λ, *the equation*

$$\sum_{j=0}^{n} a_j(t)x^{(j)}(t) + \lambda \int_J k(s, t)x(s)\, ds = y(t) \quad \text{a.e.} \qquad y \in \mathfrak{L}_1(J)$$

has precisely n *linearly independent solutions in* $A_n(J)$.

Proof. Define T to be the maximal operator corresponding to $\left(\sum_{k=0}^{n} a_j D^j, 1, 1\right)$ and J. Define K on $\mathfrak{L}_1(J)$ by

$$(Kx)(t) = \int_J k(s, t)x(s)\, ds$$

Then, by Theorems VI.3.1 and VI.3.2 and Example III.3.12, T is normally solvable, $\alpha(T) = n$, $\beta(T) = 0$, and K is strictly singular. Hence, except perhaps for isolated λ,

$$\beta(T + \lambda K) = 0 \qquad \text{and} \qquad \alpha(T + \lambda K) = \kappa(T + \lambda K) = \kappa(T) = n$$

by Theorem V.2.1.

chapter VII

THE DIRICHLET OPERATOR

In this chapter, a very small portion of the theory of partial differential operators is presented. A Dirichlet boundary value problem in a Banach space setting is considered, and the vital role played by a fundamental inequality is discussed. Few proofs are given, since the arguments are very involved and rely on some deep theorems in classical analysis. For further study and additional references, we refer the reader to Agmon, Douglis and Nirenberg [1], Bers and Schechter [1], Browder [3], [4], Hörmander [1], Friedman [1], and Sobolev [2].

VII.1 THE SOBOLEV SPACE $W^{m,p}(\Omega)$

This section is concerned with the definition and properties of the space $W^{m,p}(\Omega)$, which is referred to as a Sobolev space. The definition and discovery of its basic properties are due to Sobolev [1] and [2]. See also Friedrichs [1].

Throughout this chapter, Ω is an open subset of E^n, $n \geq 1$. The points of E^n are denoted by $x = (x_1, x_2, \ldots, x_n)$, and integration is with respect to Lebesgue measure. For $D_j = \partial/\partial x_j$ and $\alpha = (\alpha_1, \alpha_2, \ldots, \alpha_n)$, an n-tuple of nonnegative integers, we set $D^\alpha = D_1^{\alpha_1} D_2^{\alpha_2} \ldots D_n^{\alpha_n}$. The order of D^α is $|\alpha| = \sum_{i=1}^{n} \alpha_i$.

So as to motivate the definition of $W^{m,p}(\Omega)$, let us consider the one-dimensional case where Ω is an open interval I. It was shown in Theorem VI.6.2 that for $1 \leq p \leq \infty$,

$$W^{m,p}(I) = \{f \mid f^{(k)} \in \mathfrak{L}_p(I), \quad 0 \leq k \leq m, \quad f^{(m-1)} \text{ absolutely continuous}\}$$

is the domain of the maximal operator corresponding to (τ, p, p) and I, where $\tau = \sum_{k=0}^{m} a_k D^k$ with $1/a_m$ and $a_k \in \mathfrak{L}_\infty(I)$, $0 \leq k \leq m$.

Now for $f \in W^{m,p}(I)$, successive integration by parts shows that for all $\varphi \in C_0^\infty(I)$,

$$\int_I f\varphi^{(k)} = (-1)^k \int_I f^{(k)}\varphi \qquad 1 \leq k \leq m$$

In fact, $W^{m,p}(I)$ has the following characterization.

VII.1.1 Theorem. *A necessary and sufficient condition that $f \in \mathfrak{L}_p(I)$ be in $W^{m,p}(I)$, $1 \leq p \leq \infty$, is that for each integer k, $1 \leq k \leq m$, there exists a $g_k \in \mathfrak{L}_p(I)$ such that*

$$\int_I f\varphi^{(k)} = (-1)^k \int_I g_k\varphi \qquad \varphi \in C_0^\infty(I)$$

We first prove the following lemma.

VII.1.2 Lemma. *Let g be locally integrable on I. If for some positive integer k,*

$$\int_I g\varphi^{(k)} = 0 \qquad \varphi \in C_0^\infty(I)$$

then g is, almost everywhere, a polynomial of degree at most $k - 1$.

Proof. Let V be the vector space $C_0^\infty(I)$ and let V^* be the vector space of linear functionals on V. Define D to be the ordinary derivative operator mapping V into V and define the transpose D^* from V^* into V^* by $(D^*F)\varphi = FD\varphi$, $F \in V^*$, $\varphi \in V$. Let $F_g \in V^*$ be defined by

$$F_g\varphi = \int_I g\varphi \qquad \varphi \in V$$

Then, by hypothesis,

$$((D^*)^kF_g)\varphi = F_gD^k\varphi = \int_I g\varphi^{(k)} = 0 \qquad \varphi \in V$$

that is, F_g is in $\mathfrak{N}((D^*)^k)$. Assert that $F \in \mathfrak{N}((D^*)^k)$ implies that F has the representation

$$F\varphi = \int_I P\varphi \qquad \varphi \in V$$

where P is a polynomial of degree at most $k - 1$. Assume $k = 1$. Define $F_1 \in V^*$ by

$$F_1\varphi = \int_I \varphi$$

Then $\mathfrak{N}(F_1) = DV$. Indeed, suppose $\varphi = D\Phi$ for some $\Phi \in V$. Then

$$F_1\varphi = \int_I \varphi = \int_I \Phi' = \Phi(b) - \Phi(a) = 0$$

where Φ has support in compact interval $[a, b] \subset I$. Thus $DV \subset \mathfrak{N}(F_1)$. On the other hand, if φ is in $\mathfrak{N}(F_1)$ and has support in compact interval $[a, b]$, define Φ on I by

$$\Phi(x) = \int_a^x \varphi(t)\, dt$$

Then Φ is infinitely differentiable on I, and since $\int_I \varphi = 0$, Φ has support in $[a, b]$. Hence $\mathfrak{N}(F_1) \subset DV$, and therefore $\mathfrak{N}(F_1) = DV$. Given $F \in \mathfrak{N}(D^*), 0 = (D^*F)\varphi = FD\varphi$ for all $\varphi \in V$. Thus $\mathfrak{N}(F_1) = DV \subset \mathfrak{N}(F)$. Since $\dim V/\mathfrak{N}(F_1) = 1$, $F = cF_1$ for some constant c. Indeed,

$$F = F(\varphi_0)F_1$$

where φ_0 is chosen so that $F_1\varphi_0 = 1$. This means that for all $\varphi \in V$,

$$F\varphi = cF_1\varphi = \int_I c\varphi$$

Thus the assertion about $F \in \mathfrak{N}((D^*)^k)$ is proved for $k = 1$. Assume the assertion holds for $k = j$, and $0 = (D^*)^{j+1}F = (D^*)^j(D^*F)$. Then there exists a polynomial p_{j-1} of degree at most $j - 1$ such that

$$(D^*F)\varphi = \int_I p_{j-1}\varphi \qquad \varphi \in V \tag{1}$$

Choose p_j to be a polynomial so that $Dp_j = -p_{j-1}$. Then for $P_j \in V^*$ defined by

$$P_j\varphi = \int_I p_j\varphi$$

integration by parts shows that for all $\varphi \in V$,

$$(D^*P_j)\varphi = \int_I p_j\varphi' = \int_I p_{j-1}\varphi \tag{2}$$

Thus by (1) and (2), $D^*(P_j - F) = 0$. Hence, by what has been shown, there exists a constant c such that $P_j - F = cF_1$, or equivalently,

$$F\varphi = \int_I (p_j - c)\varphi$$

Since $p_j - c$ is a polynomial of degree at most j, the assertion about $F \in \mathfrak{N}((D^*)^k)$ follows by induction. In particular, since F_g is in $\mathfrak{N}((D^*)^k)$, there exists a polynomial p of degree at most $k - 1$ such that for all $\varphi \in V$,

$$0 = F_g\varphi - \int_I p\varphi = \int_I (g - p)\varphi$$

Hence $g = p$ a.e., which proves the lemma.

Proof of the theorem. The necessity was proved in the discussion preceding the theorem. Assume the existence of the g_k as in the theorem. Successive integration by parts implies the existence of a G_k such that $G_k^{(k-1)}$ is absolutely continuous, $G_k^{(k)} = g_k$ a.e., and

$$\int_I g_k\varphi = (-1)^k \int_I G_k\varphi^{(k)} \qquad \varphi \in C_0^\infty(I)$$

Hence for all $\varphi \in C_0^\infty(I)$,

$$\int_I (f - G_k)\varphi^{(k)} = 0 \qquad 1 \leq k \leq m$$

Thus $f - G_k$ is a polynomial, almost everywhere, of degree at most $k - 1$ by Lemma VII.1.2. In particular, there exists a constant c such that $f^{(m-1)}$ is $G_m^{(m-1)} + c$ a.e. and is absolutely continuous, and $f^{(k)} = G_k^{(k)} = g_k$ as elements of $\mathfrak{L}_p(I)$, $1 \leq k \leq m$. Hence f is in $W^{m,p}(I)$.

Based on Theorem VII.1.1, we obtain the following characterization of $W^{m,p}(I)$ for $1 < p \leq \infty$.

Let D be the ordinary derivative considered as a map from $C_0^\infty(I) \subset \mathfrak{L}_{p'}(I)$ into $\mathfrak{L}_{p'}(I)$, where p' is conjugate to p and $1 < p \leq \infty$. Let f and g_k be as in Theorem VII.1.1 considered as elements of the conjugate space $\mathfrak{L}_p(I) = \mathfrak{L}'_{p'}(I)$. Then

$$f(D^k\varphi) = (-1)^k g_k(\varphi) \qquad \varphi \in \mathfrak{D}(D)$$

Thus f is in $W^{m,p}(I)$ if and only if f is in $\mathfrak{D}((D^k)')$ and $(D^k)'f = (-1)^k g_k$.

In light of the above discussion, the Sobolev space is defined as follows.

VII.1.3 Definition. *Let Ω be an open subset of E^n. Given a nonnegative integer m, define $W^{m,p}(\Omega)$, $1 \leq p \leq \infty$, as the linear space of those $f \in \mathfrak{L}_p(\Omega)$ having the property that for $|\alpha| \leq m$, there exists a $g_\alpha \in \mathfrak{L}_p(\Omega)$ such that*

$$(-1)^{|\alpha|} \int_\Omega f D^\alpha \varphi = \int_\Omega g_\alpha \varphi \qquad \varphi \in C_0^\infty(\Omega)$$

The norm on $W^{m,p}(\Omega)$ is given by

$$\|f\|_{m,p} = \Big(\sum_{|\alpha| \leq m} \|g_\alpha\|_p^p \Big)^{1/p}$$

The generalized α-derivative of f is defined to be g_α which is unique by Theorem 0.10. *We write $D^\alpha f = g_\alpha$.*

VII.1.4 Remarks

i. For $f \in C^m(\Omega) \cap W^{m,p}(\Omega)$, integration by parts shows that the generalized α-derivative of f, $|\alpha| \leq m$, coincides with the α-derivative of f in the classical sense.

ii. For $1 < p \leq \infty$, the definition of the space $W^{m,p}(\Omega)$ is equivalent to the following definition.
Consider D^α as a linear map from $C_0^\infty(\Omega) \subset \mathfrak{L}_{p'}(\Omega)$ into $\mathfrak{L}_{p'}(\Omega)$, p' conjugate to p. Then

$$W^{m,p}(\Omega) = \{f \mid f \in \mathfrak{D}((D^\alpha)') \subset \mathfrak{L}_p(\Omega),\ |\alpha| \leq m\}$$

where $(D^\alpha)'$ is the conjugate of D^α. Moreover,

$$(-1)^{|\alpha|}(D^\alpha)'f = g_\alpha$$

where g_α is the generalized α-derivative of f. Thus the generalized α-derivative is a conjugate operator.

iii. Meyers and Serrin [1] proved that for $1 \leq p < \infty$ and Ω any open set in E^n, $C^\infty(\Omega) \cap W^{m,p}(\Omega)$ is dense in $W^{m,p}(\Omega)$ with respect to the $W^{m,p}(\Omega)$ norm.

For the remainder of the chapter, we assume $1 < p < \infty$.

VII.1.5 Theorem. *$W^{m,p}(\Omega)$ is reflexive.*

Proof. Using the fact that D^α is closed (since it is a conjugate operator), the proof is similar to the proof of Remark VI.6.4 that $W^{m,p}(I)$ is reflexive.

The Sobolev spaces for certain bounded Ω have the very important property that the identity map from $W^{j,p}(\Omega)$ into $W^{k,p}(\Omega)$, $0 \leq k < j$, is compact. To be precise, we state the following result. For more general Ω, the reader is referred to Browder [3], Lemma 5. One of the original versions of the theorem appears in Sobolev [2], Theorem 2, page 84.

VII.1.6 Definition. *The boundary $\partial\Omega$ of Ω is called* **smooth** *if for each $x \in \partial\Omega$ there exists an open set $\mathcal{O}_x$ about x and a mapping φ_x of $\mathcal{O}_x$ onto an open ball about the origin in E^n such that*

i. φ_x *is* 1-1 *and both* φ_x and φ_x^{-1} *are infinitely differentiable.*
ii. $\varphi_x(\mathcal{O}_x \cap \partial\Omega) = \varphi_x(\mathcal{O}_x) \cap \{x \mid x \in E^n, x_1 = 0\}$

VII.1.7 Theorem. *If Ω is bounded with smooth boundary, no point of which is interior to the closure of Ω, then the identity map from $W^{j,p}(\Omega)$ into $W^{k,p}(\Omega)$, $0 \leq k < j$, is compact.*

VII.2 A PRIORI ESTIMATES AND DIFFERENTIAL OPERATORS

We have seen in Secs. VI.6 to VI.8 the vital role played by the a priori estimate given in VI.6.2, namely, the estimate which shows that the norm on $W^{m,p}(I)$ is dominated by the T-norm for certain differential operators T. A similar estimate holds for certain partial differential operators T defined on a subspace of $W^{m,p}(\Omega)$. However, the estimate is very difficult to establish. See, for example, the papers of Browder [3] and Agmon, Douglis, and Nirenberg [1]. A number of references pertaining to the estimate are also contained in these papers.

The following result shows how the a priori estimate, sometimes referred to as a coerciveness inequality, is related to certain properties of the operator. The theorem is motivated by the fact that $W^{m,p}(\Omega)$ is complete, the identity map from $W^{m,p}(\Omega)$ into $W^{o,p}(\Omega) = \mathcal{L}_p(\Omega)$ is compact (for Ω as in Theorem VII.1.7), and $\|f\|_{m,p} \geq \|D^\alpha f\|_p$ for all $f \in W^{m,p}(\Omega)$, $|\alpha| \leq m$.

VII.2.1 Theorem. *Let T be a linear operator with domain $\mathfrak{D}$ in Banach space X and range in Banach space Y. Let $\mathfrak{D}_1$ be $\mathfrak{D}$ with norm $\| \quad \|_1$ such that $\mathfrak{D}_1$ is complete and the identity map from $\mathfrak{D}_1$ onto $\mathfrak{D}$ with the T-norm is bounded. Then*

i. T is closed if and only if there exists a number K such that for all $f \in \mathfrak{D}$,

$$(1) \qquad \|f\|_1 \leq K(\|f\| + \|Tf\|)$$

ii. If (1) *holds and the identity map from $\mathfrak{D}_1$ into X is compact, then T is normally solvable and has finite kernel index.*

Proof of (i). Let $\mathfrak{D}_T$ be $\mathfrak{D}$ with the T-norm and let I be the identity map from $\mathfrak{D}_1$ onto $\mathfrak{D}_T$. It is easy to see that T is closed if and only if $\mathfrak{D}_T$ is complete. Since $\mathfrak{D}_1$ is complete and I is bounded, we know from the open-mapping theorem that the completeness of $\mathfrak{D}_T$ implies that I has a bounded inverse or, equivalently, (1) holds. Conversely, if (1) holds, then I is an isomorphism and therefore $\mathfrak{D}_T$ is complete. Thus (*i*) is proved.

Proof of (ii). Let S be the 1-ball of $\mathfrak{N}(T)$ considered as a subspace of $\mathfrak{D}_1$ and let I_1 be the identity map from $\mathfrak{D}_1$ into X. By hypothesis, I_1 is compact, and therefore S is totally bounded in X. Since each $f \in S$ satisfies $\|f\|_1 \leq K(\|f\| + \|Tf\|) = K\|f\|$, the total boundedness of S in X implies the total boundedness of S in $\mathfrak{D}_1$. Hence $\mathfrak{N}(T)$ is finite-dimensional. Consequently, $X = \mathfrak{N}(T) \oplus M$ where M is a closed subspace of X. Let T_M be the restriction of T to $M \cap \mathfrak{D}$. Then T_M is 1-1 and closed and $\mathfrak{R}(T_M) = \mathfrak{R}(T)$. Assume $\mathfrak{R}(T)$ is not closed. Then by Lemma IV.1.1, T_M does not have a bounded inverse. Thus there exists a sequence $\{m_k\}$ in M such that $\|m_k\| = 1$ and $Tm_k \to 0$. Hence $\{m_k\}$ is bounded in $\mathfrak{D}_1$ by (1). Since I_1 is compact and M is closed in X, there exists a subsequence $\{m_{k'}\}$ of $\{m_k\}$ which converges in X to some $m \in M$. Now T_M is closed and $T_M m_{k'} \to 0$. Thus m is in $\mathfrak{D}(T_M)$, and $T_M m = 0$. But this is impossible, since $\|m\| = 1$ and T_M is 1-1. Hence $\mathfrak{R}(T)$ is closed.

In view of the above theorem, we shall study certain important partial differential operators for which the a priori estimate holds. In this way additional examples of normally solvable and even Fredholm operators are obtained.

VII.2.2 Definition. *Let τ be the partial differential expression*

$$\tau = \sum_{|\alpha| \leq m} a_\alpha D^\alpha$$

where the a_α are complex-valued functions defined on a set $S \subset E^n$. We call τ ***elliptic*** *on S if for every $x \in S$ and every $\xi \neq 0$ in E^n,*

$$\tau_m(x, \xi) = \sum_{|\alpha| = m} a_\alpha(x)\xi^\alpha \neq 0 \qquad \xi^\alpha = \xi_1^{\alpha_1}\xi_2^{\alpha_2} \cdots \xi_n^{\alpha_n}$$

τ_m is called the ***characteristic form*** *of τ and m is the order of τ.*

The Laplacian $\tau = D_1^2 + D_2^2 + \cdots D_n^2$ is elliptic and of order 2 on E^n since $\tau_2(x, \xi) = \xi_1^2 + \xi_2^2 + \cdots + \xi_n^2$.

The Cauchy-Riemann expression $\tau = D_1 + iD_2$ is elliptic and of order one on E^2, since $\tau_1(x, \xi) = \xi_1 + i\xi_2$.

Any ordinary differential expression with complex-valued coefficients defined on an interval I and with nonvanishing leading coefficient is elliptic on I.

VII.2.3 Remark. Lopatinski [1] proved that if τ is elliptic on $\Omega \subset E^n$, $n > 2$, then the order of τ is even. A detailed proof appears in Lions [1], page 65. The expression $\tau = D_1 + iD_2$ shows that this is not true for $n = 2$.

The next two theorems, which show how closely the ellipticity of τ and the a priori estimate are related, are due to Gårding [1] and Schechter [1], respectively.

VII.2.4 Theorem. *Let Ω be any open set and let τ be an mth-order partial differential expression with constant coefficients. A necessary and sufficient condition that there exist a constant K such that*

$$\|f\|_{m,2} \leq K(\|f\|_2 + \|\tau f\|_2) \qquad \text{for all } f \in C_0^\infty(\Omega)$$

is that τ be elliptic.

VII.2.5 Theorem. *Let Ω be bounded in E^n, $n > 2$, with smooth boundary $\partial\Omega$ and let τ be an mth-order partial differential expression with continuous complex-valued coefficients on $\bar{\Omega}$. Then there exists a constant K such that*

$$\|f\|_{m,2} \leq K(\|f\|_2 + \|\tau f\|_2)$$

for all $f \in \{u \mid u \in C^\infty(\bar{\Omega}), D^\beta u = 0$ on $\partial\Omega$, $|\beta| < r\}$ if and only if τ is elliptic on $\bar{\Omega}$ and $m \leq 2r$.

The operator we consider next is one which is determined by τ and by the boundary conditions $D^\beta u = 0$ on $\partial\Omega$, $|\beta| < m$, where τ is of order $2m$. There are two reasons for considering precisely m boundary conditions. The first is that the a priori estimate does not hold if there are less than m boundary conditions, as seen in Theorem VII.2.5. The second reason is that the operator and its conjugate are nicely related if there are m boundary conditions. This is seen in Theorem VII.2.13.

For simplicity, the following assumptions are made for the remainder of the chapter.

i. Ω is bounded with smooth boundary $\partial\Omega$, no point of which is interior to the closure of Ω.

ii. $\tau = \sum_{|\alpha| \leq 2m} a_\alpha D^\alpha$ is elliptic on the closure $\bar{\Omega}$ of Ω (by Remark VII.2.3, τ is of even order for $n > 2$). Each a_α is in $C^\infty(\bar{\Omega})$; that is, a_α is a restriction to $\bar{\Omega}$ of some function in $C^\infty(\mathcal{O})$, where $\mathcal{O}$ is an open set in E^n which contains $\bar{\Omega}$.

VII.2.6 Definition. *Let $W_0^{m,p}(\Omega)$ be the closure in $W^{m,p}(\Omega)$ of $C_0^\infty(\Omega)$ and let T_p be the linear map from a subspace of $\mathfrak{L}_p(\Omega)$ into $\mathfrak{L}_p(\Omega)$ defined by*

$$\mathfrak{D}(T_p) = \{u \mid u \in W_0^{m,p}(\Omega) \cap W^{2m,p}(\Omega)\}$$

$$T_p u = \tau u$$

T_p is called a ***Dirichlet operator.***

VII.2.7 Remarks

i. $\mathfrak{R}(T_p)$ is indeed a subspace of $\mathfrak{L}_p(\Omega)$, since a_α is continuous on $\bar{\Omega}$ and $D^\alpha u$ is in $\mathfrak{L}_p(\Omega)$, $|\alpha| \leq 2m$, $u \in W^{2m,p}(\Omega)$.

ii. If u is in $C^\infty(E^n) \cap \mathfrak{D}(T_p)$, then $D^\beta u = 0$ on $\partial\Omega$, $|\beta| < m$, in the classical sense; that is, $(\partial^{j-1}/\partial\nu^{j-1})u = 0$, $j = 1, 2, \ldots, m$, where $\partial/\partial\nu$ is the normal derivative.

iii. For a characterization of $\mathfrak{D}(T_p)$, see Browder [3], Lemma 9.

The next theorem is a special case of a result due to Agmon, Douglis, and Nirenberg [1], and Browder [3].

VII.2.8 Theorem. *For $n > 2$, there exists a constant K_p such that for all $u \in \mathfrak{D}(T_p)$,*

$$\|u\|_{2m,p}^p \leq K_p(\|u\|_p^p + \|Tu\|_p^p)$$

VII.2.9 Corollary. *For $n > 2$, T_p is a Fredholm operator.*

The proof that $\beta(T_p) < \infty$ is quite involved. See, for example, Browder [3], Theorem 6. The idea is to construct a linear operator S which maps $\mathfrak{L}_p(\Omega)$ into $W^{m,p}(\Omega)$ such that $T_pS = I + K$, where K is compact. Since $\mathfrak{R}(T_pS) \subset \mathfrak{R}(T_p)$, one obtains

$$\beta(T_p) \leq \beta(T_pS) = \beta(I + K) < \infty$$

We shall only prove that T_p is normally solvable and has finite kernel index.

Proof. Let $\mathfrak{D}_p$ be $\mathfrak{D}(T_p)$ with norm $\| \quad \|_{2m,p}$. Since $W_0^{m,p}(\Omega) \cap W^{2m,p}(\Omega)$ is a closed subspace of Banach space $W^{2m,p}(\Omega)$, $\mathfrak{D}_p$ is complete. Moreover,

$$\|T_pu\|_p \leq \Big(\sum_{|\alpha|\leq 2m} \sup_{x\epsilon\Omega} |a_\alpha(x)|\Big) \|u\|_{2m,p}$$

and the identity map from $\mathfrak{D}_p$ onto $\mathfrak{D}(T_p)$ with the T_p-norm is continuous. Hence T_p is normally solvable and has finite kernel index by Theorem VII.2.1.

When $n = 2$, an additional condition on τ is required in order for the above theorem and its corollary to hold. This condition is the following.

VII.2.10 Condition on τ_m. *At each $x \in \partial\Omega$, with ν a normal to $\partial\Omega$ at x and t any unit tangent vector to $\partial\Omega$ at x, the polynomial*

$$\tau_m(x, t + \lambda\nu) = \sum_{|\alpha| = 2m} a_\alpha(x)(t + \lambda\nu)^\alpha$$

in the complex variable λ has exactly m zeros with positive imaginary parts. (The remaining m zeros must have negative imaginary parts, since ellipticity implies there are no real zeros.)

Lopatinski [1] showed that the condition holds for $n > 2$. For general n, the condition holds if each a_α, $|\alpha| = 2m$, is real-valued.

VII.2.11 Definition. *Let $T_{p',*}$ be the linear operator defined from a subspace of $\mathfrak{L}_{p'}(\Omega)$ into $\mathfrak{L}_{p'}(\Omega)$, p' conjugate to p, by*

$$\mathfrak{D}(T_{p',*}) = \{v \mid v \in W^{2m,p'}(\Omega),\ D^\beta v = 0 \text{ on } \partial\Omega,\ |\beta| < m\}$$

$$T_{p',*}v = \sum_{|\alpha| \leq 2m} (-1)^{|\alpha|} D^\alpha(a_\alpha v)$$

Note that $a_\alpha v$ is in $W^{2m,p'}(\Omega)$, $|\alpha| \leq 2m$, since

$$\left|\int a_\alpha v D\varphi\right| \leq \Big(\sum_{|\alpha| \leq 2m} \sup_{x \in \Omega} |a_\alpha(x)| \Big) \|D^\alpha v\|_{p'} \|\varphi\|_p \qquad \varphi \in C_0^\infty(\Omega)$$

that is, $a_\alpha v D^\alpha$ is continuous on $C_0^\infty(\Omega) \subset \mathfrak{L}_p(\Omega)$.

VII.2.12 Remark. In the above definition, take $\alpha = (1, 0, \ldots, 0)$ and $a = a_\alpha$. Then

$$\text{(1)} \qquad D_1(av)\varphi = -\int_\Omega avD_1\varphi = -\int_\Omega vD_1(a\varphi) + \int_\Omega (D_1a)v\varphi$$

$$= (D_1v)(a\varphi) + ((D_1a)v)\varphi$$

If we agree to let $(bg)\varphi = g(b\varphi)$, where g is any linear functional on $C_0^\infty(\Omega)$ and b is in $C^\infty(\Omega)$, then (1) shows that

$$D_1(av) = aD_1v + (D_1a)v$$

It follows by differentiating successively that

$$T_{p',*}v = \tau^* v = \sum_{|\alpha| \le 2m} a_\alpha^* D^\alpha v$$

where τ^* is elliptic on $\bar{\Omega}$ and the a_α^* are in $C^\infty(\Omega)$.

The next two theorems are special cases of Theorems 5 and 9, respectively, in Browder [3].

VII.2.13 Theorem. *For $p = 2$, $T'_2 = T_{2,*}$.*

For $p \neq 2$, the following additional hypotheses are used by Browder to determine T'_p.

VII.2.14 Theorem. *Suppose that τ and τ^* have the following local properties. Given a connected open set $G \subset E^n$, every v and w in $C^{2m}(G)$ which satisfy $\tau v = 0$ and $\tau^* w = 0$ and which vanish on an open subset of G must vanish identically on G. Under these conditions, $T'_p = T_{p',*}$.*

VII.2.15 Remark. While we did not prove in Corollary VII.2.9 that $\beta(T_p)$ is finite, we did show that T_p is normally solvable and has finite kernel index. Now, with Theorem VII.2.14 we have

$$\beta(T_2) = \alpha(T'_2) = \alpha(T_{2,*}) < \infty$$

since τ^* is of the same form as τ.

We describe briefly the operator determined by general boundary conditions. For a more complete description, the reader is referred to Agmon, Douglis, and Nirenberg [1]; Browder [1]; and Schechter [2] and [3].

VII.2.16 Definition. *Let $B_j = \sum_{|\alpha| \le m_j} b_\alpha{}^j D^\alpha$, $j = 1, 2, \ldots, m$, be a differential expression of order $m_j < 2m$ with coefficients in $C^{2m}(\partial\Omega)$; that is, $b_\alpha{}^j$ is the restriction of some function in $C^{2m}(\mathcal{O})$, where $\mathcal{O}$ is an open set in E^n containing $\partial\Omega$. Define T_p as a map from a subspace of $\mathcal{L}_p(\Omega)$ into $\mathcal{L}_p(\Omega)$ by:*

$\mathfrak{D}(T_p)$ is the closure in $W^{2m,p}(\Omega)$ of those $u \in C^{2m}(\bar{\Omega})$ for which $B_j u = 0$ on $\partial\Omega$, $1 \le j \le m$.

$$T_p f = \tau f$$

VII.2.17 Definition. *Let the characteristic form τ_{2m} of τ satisfy condition* VII.2.8. *We say that the set $\{B_1, B_2, \ldots, B_n\}$ satisfies the* **complementing condition** *for τ if at each point $x \in \partial\Omega$, the polynomials $\sum_{|\alpha| = m_j} b_\alpha{}^j(x)(t + \lambda\nu)^\alpha$ in the complex variable λ are linearly independent* *modulo* $(\lambda - \lambda_1)(\lambda - \lambda_2) \ldots (\lambda - \lambda_m)$ *for each* $t \neq 0$ *tangent to* $\partial\Omega$ *at* x *and each* $\nu \neq 0$ *normal to* $\partial\Omega$ *at* x, *where* $\lambda_1, \lambda_2, \ldots, \lambda_m$ *are the zeros of* $\tau_{2m}(x, t + \lambda\nu)$ *with positive imaginary parts.*

VII.2.18 **Theorem.** *Let the characteristic form of* τ *satisfy condition* VII.2.10. *Then* T_p *is closed, or equivalently, there exists a constant* K *such that for all* $f \in \mathfrak{D}(T_p)$,

$$\|f\|_{2m,p} \leq K(\|f\|_p + \|Tf\|_p)$$

if and only if $\{B_1, B_2, \ldots, B_m\}$ *satisfies the complementing condition for* τ. *In this case,* T_p *is a Fredholm operator.*

The proof that T_p is normally solvable with finite kernel index is the same as in Corollary VII.2.9.

Using perturbation theory, more general boundary operators were considered by Beals [1] and Freeman [1].

BIBLIOGRAPHY

Agmon, S., A. Douglis, and L. Nirenberg

[1] Estimates near the Boundary for Solutions of Elliptic Partial Differential Equations Satisfying General Boundary Conditions I, *Commun. Pure Appl. Math.*, vol. 12, no. 4, pp. 623–727, 1959.

Akhiezer, N. I., and I. M. Glazman

[1] "Theory of Linear Operators in Hilbert Space," vols. I and II, Frederick Ungar Publishing Co., New York, 1962, 1963.

Atiyah, M. F., and I. M. Singer

[1] The Index of Elliptic Operators on Compact Manifolds, *Bull. Am. Math. Soc.*, vol. 69, no. 3, pp. 422–432, 1963.

Balslev, E., and T. W. Gamelin

[1] The Essential Spectrum of a Class of Ordinary Differential Operators, *Pacific J. Math.*, vol. 14, no. 3, pp. 755–776, 1964.

Banach, S.

[1] "Théorie des opérations linéaires," Monografje Matematyczne, Warsaw, 1932.

Beals, R. W.

[1] Non-local Boundary Value Problems for Elliptic Operators, Thesis, Yale University, 1964.

[2] A Note on the Adjoint of a Perturbed Operator, *Bull. Am. Math. Soc.*, vol. 70, no. 2, pp. 314–315, 1964.

Berberian, S. K.

[1] "Introduction to Hilbert Space," Oxford University Press, Fair Lawn, N.J., 1961.

Bers, L., and M. Schechter

[1] "Elliptic Equations," Partial Differential Equations, Proc. summer seminar, Boulder, Colo., 1957, pp. 131–299, Interscience Publishers, Inc., New York, 1964.

Birman, M. S.

[1] On the Spectrum of Singular Boundary Value Problems (Russian), *Mat. Sb.*, Ser. 2, vol. 55, pp. 125–174, 1961.

Borsuk, K.

[1] Drei Sätze über die n-dimensionale euklidische Sphäre, *Fund. Math.*, vol. 20, pp. 177–190, 1933.

Bourbaki, N.

[1] "Espaces vectoriels topologiques," "Eléments de mathématique," livre V, *Actualités Sci. Ind.*, 1229, Hermann & Cie, Paris, 1955.

Browder, F. E.

[1] Estimates and Existence Theorems for Elliptic Boundary Value Problems, *Proc. Natl. Acad. Sci. U.S.*, vol. 45, no. 3, pp. 365–372, 1959.

[2] Functional Analysis and Partial Differential Equations I, *Math. Ann.*, vol. 138, pp. 55–79, 1959.

[3] On the Spectral Theory of Elliptic Differential Operators I, *Math. Ann.*, vol. 142, pp. 22–130, 1961.

[4] Functional Analysis and Partial Differential Equations II, *Math. Ann.*, vol. 145, pp. 81–226, 1962.

Day, M. M.

[1] On the Basis Problem in Normed Spaces, *Proc. Am. Math. Soc.*, vol. 13, no. 4 pp. 655–658, 1962.

Dieudonné, J.

[1] La dualité dans les espaces vectoriels topologiques, *Ann. Sci. Ecole Norm. Super.* Ser. 3, vol. 59, pp. 107–139, 1942.

Dunford, N., and J. T. Schwartz

[1] "Linear Operators," parts I and II, Interscience Publishers, Inc., New York, 1958, 1963.

Feldman, I. A., I. C. Gokhberg, and A. S. Markus

[1] Normally Solvable Operators and Ideals Associated with Them (Russian), Izv. Moldavsk Filiala (Akad. Nauk) SSSR vol. 10 no. 76, pp. 51–69, 1960. Translated into English by T. W. Gamelin, Math. Dept. Rep., University of California, Berkeley, 1963.

Fredholm, I.

[1] Sur une classe d'equations fonctionelles, *Acta Math.*, vol. 27, pp. 365–390, 1903.

Freeman, R. S.

[1] Closed Operators and Their Adjoints Associated with Elliptic Differential Operators, to appear.

Friedman, A.

[1] "Generalized Functions and Partial Differential Equations," Prentice-Hall, Inc., Englewood Cliffs, N.J., 1963.

Friedrichs, K. O.

[1] On Differential Operators in Hilbert Space, *Am. J. Math.*, vol. 61, pp. 523–544, 1939.

Gårding, L.

[1] Dirichlet's Problem for Linear Partial Differential Equations, *Math. Scand.*, vol. 1, pp. 55–72, 1953.

Gokhberg, I. C., and M. G. Krein

[1] Fundamental Theorems on Deficiency Numbers, Root Numbers, and Indices of Linear Operators (Russian), *Usp. Mat. Nauk*, 12, pp. 43–118, 1957. Translated in *Am. Math. Soc. Transls.*, ser. 2, vol. 13.

Goldberg, S.

[1] Linear Operators and Their Conjugates, *Pacific J. Math.*, vol. 9, no. 1, pp. 69–79, 1959.

[2] Closed Linear Operators and Associated Continuous Linear Operators, *Pacific J. Math.*, vol. 12, no. 1, pp. 183–186, 1962.

Goldberg, S., and A. H. Kruse

[1] The Existence of Compact Linear Maps between Banach Spaces, *Proc. Am. Math. Soc.*, vol. 13, no. 5, pp. 808–811, 1962.

Goldberg, S., and E. O. Thorp

[1] On Some Open Questions Concerning Strictly Singular Operators, *Proc. Am. Math. Soc.*, vol. 14, no. 2, pp. 334–336, 1963.

Goldman, M. A.

[1] On the Stability of the Property of Normal Solvability of Linear Equations (Russian), *Dokl. Akad. Nauk SSSR,* (N. S.) vol. 100, pp. 201–204, 1955.

Halmos, P. R.

[1] "Introduction to Hilbert Space," Chelsea Publishing Company, New York, 1951.

Halperin, I.

[1] Closures and Adjoints of Linear Differential Operators, *Ann. Math.* (2) 38, pp. 880–919, 1937.

Hörmander, L.

[1] "Linear Partial Differential Operators," vol. 116, "The Essentials of Mathematics," Academic Press, Inc., New York, 1963.

James, R. C.

[1] A Non-reflexive Banach Space Isometric with Its Second Conjugate Space, *Proc. Natl. Acad. Sci. U.S.*, vol. 37, pp. 174–177, 1951.

Kato, T.

[1] Perturbation Theory for Nullity, Deficiency and Other Quantities of Linear Operators, *J. d'Analyse Math.*, vol. 6, pp. 273–322, 1958.

Kemp, R. R. D.

[1] On a Class of Singular Differential Operators, *Can. J. Math.*, vol. 13, pp. 316–330, 1961.

Krein, M. G., M. A. Krasnoselskii, and D. P. Milman

[1] On the Defect Numbers of Linear Operators in a Banach Space and on Some Geometrical Questions (Russian), *Sb. Tr. Inst. Mat. Akad. Nauk Ukr. SSR*, no. 11, pp. 97–112, 1948.

Krishnamurthy, V.

[1] On the State of a Linear Operator and Its Adjoint, *Math. Ann.*, vol. 141, pp. 153–160, 1960.

Lacey, E.

[1] Generalizations of Compact Operators in Locally Convex Topological Linear Spaces, Thesis, New Mexico State University, 1963.

Lacey, E., and R. J. Whitley

[1] Conditions under Which All the Bounded Linear Maps Are Compact, *Math. Ann.*, vol. 158, pp. 1–5, 1965.

Lions, J. L.

[1] Lectures on Elliptic Partial Differential Equations, Tata Institute lecture notes, Bombay, 1957.

Loomis, L. H.

[1] "An Introduction to Abstract Harmonic Analysis," D. Van Nostrand Company, Inc., Princeton, N. J., 1953.

Lopatinski, Ya. B.

[1] On a Method of Reducing Boundary Problems for a System of Differential Equations of Elliptic Type to Regular Equations (Russian), *Usp. Mat. Nauk*, vol. 13, pp. 29–89, 1958.

McShane, E. J.

[1] "Integration," Princeton University Press, Princeton, N. J., 1947.

McShane, E. J., and T. Botts

[1] "Real Analysis," D. Van Nostrand Company, Inc., Princeton, N. J., 1959.

Meyers, N. G., and J. Serrin

[1] H = W., *Proc. Natl. Acad. Sci. U.S.*, 51, pp. 1055–1056, 1964.

Naimark, M. A., see Neumark

Neumark (Naimark), M. A.

[1] "Lineare differentialoperatoren," Akademie Verlag, Berlin, 1960.

Pelczynski, A.

[1] Projections in Certain Banach Spaces, *Studia Math.*, vol. 19, pp. 209–228, 1960.

[2] On Strictly Singular and Strictly Cosingular Operators, vols. I and II, *Bull. Acad. Polon. Sci.* III, Series Math., Astron., Phys., vol. 13, no. 1, 1965.

Robertson, A. P., and W. J. Robertson

[1] "Topological Vector Spaces," Cambridge University Press, London, 1964.

Rota, G. C.

[1] Extension Theory of Differential Operators, *Comm. Pure Appl. Math.*, vol. 11, pp. 23–65, 1958.

[2] On the Spectra of Singular Boundary Value Problems, *J. Math. Mech.*, vol. 10, pp. 83–90, 1961.

Schechter, M.

[1] On Estimating Elliptic Partial Differential Operators in the L_2 Norm, *Am. J. Math.*, vol. 79, pp. 431–443, 1957.

[2] Integral Inequalities for Partial Differential Operators and Functions Satisfying General Boundary Conditions, *Comm. Pure. Appl. Math.*, vol. 12, pp. 37–66, 1959.

[3] General Boundary Value Problems for Elliptic Partial Differential Equations, *Comm. Pure Appl. Math.*, vol. 12, pp. 457–486, 1959.

Schwartz, L.

[1] "Théorie des distributions," Vols. I and II, Actualités Sci. Ind., 1091 and 1122, Hermann & Cie., Paris, 1951.

Sobolev, S. L.

[1] On a Theorem of Functional Analysis, *Mat. Sb.*, (N. S.) vol. 4, pp. 471–497, 1938.

[2] "Applications of Functional Analysis in Mathematical Physics" (Russian), Leningrad, 1950. Translated in *Am. Math. Soc. Transls.* vol. 7, 1963.

Stone, M. H.

[1] "Linear Transformations in Hilbert Spaces and Their Applications to Analysis," American Mathematical Society (American Mathematical Society Colloquium Publications, vol. 15) New York, 1932.

Sz.-Nagy, B.

[1] Perturbations des transformation linéaires fermées, *Acta Sci. Math.*, Szeged, vol. 14, pp. 125–137, 1951.

Taylor, A. E.

[1] "Introduction to Functional Analysis," John Wiley & Sons, Inc., New York, 1958.

Visik, M. I.

[1] On General Boundary Problems for Elliptic Differential Equations (Russian), *Tr. Mosk. Mat. Obsc.*, vol. I, pp. 187–246, 1952. Translated in *Am. Math. Soc. Transls.*, Ser. 2, vol. 24, pp. 107–172, 1963.

Whitley, R. J.

[1] Strictly Singular Operators and Their Conjugates, *Trans. Am. Math. Soc.*, vol. 13, no. 2, pp. 252–261, 1964.

Zaanen, A. C.

[1] "Linear analysis," North-Holland Publishing Co., Amsterdam, 1953.

INDEX

A CATALOGUE OF SELECTED DOVER BOOKS IN ALL FIELDS OF INTEREST

A CATALOGUE OF SELECTED DOVER BOOKS IN ALL FIELDS OF INTEREST

DRAWINGS OF WILLIAM BLAKE, William Blake. 92 plates from Book of Job, *Divine Comedy, Paradise Lost,* visionary heads, mythological figures, Laocoon, etc. Selection, introduction, commentary by Sir Geoffrey Keynes. 178pp. 8⅛ x 11. 22303-5 Pa. $4.00

ENGRAVINGS OF HOGARTH, William Hogarth. 101 of Hogarth's greatest works: *Rake's Progress, Harlot's Progress, Illustrations for Hudibras, Before and After, Beer Street and Gin Lane,* many more. Full commentary. 256pp. 11 x 13¾. 22479-1 Pa. $12.95

DAUMIER: 120 GREAT LITHOGRAPHS, Honore Daumier. Wide-ranging collection of lithographs by the greatest caricaturist of the 19th century. Concentrates on eternally popular series on lawyers, on married life, on liberated women, etc. Selection, introduction, and notes on plates by Charles F. Ramus. Total of 158pp. 9⅜ x 12¼. 23512-2 Pa. $6.00

DRAWINGS OF MUCHA, Alphonse Maria Mucha. Work reveals draftsman of highest caliber: studies for famous posters and paintings, renderings for book illustrations and ads, etc. 70 works, 9 in color; including 6 items not drawings. Introduction. List of illustrations. 72pp. 9⅜ x 12¼. (Available in U.S. only) 23672-2 Pa. $4.00

GIOVANNI BATTISTA PIRANESI: DRAWINGS IN THE PIERPONT MORGAN LIBRARY, Giovanni Battista Piranesi. For first time ever all of Morgan Library's collection, world's largest. 167 illustrations of rare Piranesi drawings—archeological, architectural, decorative and visionary. Essay, detailed list of drawings, chronology, captions. Edited by Felice Stampfle. 144pp. 9⅜ x 12¼. 23714-1 Pa. $7.50

NEW YORK ETCHINGS (1905-1949), John Sloan. All of important American artist's N.Y. life etchings. 67 works include some of his best art; also lively historical record—Greenwich Village, tenement scenes. Edited by Sloan's widow. Introduction and captions. 79pp. 8⅜ x 11¼. 23651-X Pa. $4.00

CHINESE PAINTING AND CALLIGRAPHY: A PICTORIAL SURVEY, Wan-go Weng. 69 fine examples from John M. Crawford's matchless private collection: landscapes, birds, flowers, human figures, etc., plus calligraphy. Every basic form included: hanging scrolls, handscrolls, album leaves, fans, etc. 109 illustrations. Introduction. Captions. 192pp. 8⅞ x 11¾. 23707-9 Pa. $7.95

DRAWINGS OF REMBRANDT, edited by Seymour Slive. Updated Lippmann, Hofstede de Groot edition, with definitive scholarly apparatus. All portraits, biblical sketches, landscapes, nudes, Oriental figures, classical studies, together with selection of work by followers. 550 illustrations. Total of 630pp. 9⅛ x 12¼. 21485-0, 21486-9 Pa., Two-vol. set $15.00

THE DISASTERS OF WAR, Francisco Goya. 83 etchings record horrors of Napoleonic wars in Spain and war in general. Reprint of 1st edition, plus 3 additional plates. Introduction by Philip Hofer. 97pp. 9⅜ x 8¼. 21872-4 Pa. $4.00

THE ANATOMY OF THE HORSE, George Stubbs. Often considered the great masterpiece of animal anatomy. Full reproduction of 1766 edition, plus prospectus; original text and modernized text. 36 plates. Introduction by Eleanor Garvey. 121pp. 11 x 14¾. 23402-9 Pa. $6.00

BRIDGMAN'S LIFE DRAWING, George B. Bridgman. More than 500 illustrative drawings and text teach you to abstract the body into its major masses, use light and shade, proportion; as well as specific areas of anatomy, of which Bridgman is master. 192pp. 6½ x 9¼. (Available in U.S. only) 22710-3 Pa. $3.50

ART NOUVEAU DESIGNS IN COLOR, Alphonse Mucha, Maurice Verneuil, Georges Auriol. Full-color reproduction of *Combinaisons ornementales* (c. 1900) by Art Nouveau masters. Floral, animal, geometric, interlacings, swashes—borders, frames, spots—all incredibly beautiful. 60 plates, hundreds of designs. 9⅜ x 8-1/16. 22885-1 Pa. $4.00

FULL-COLOR FLORAL DESIGNS IN THE ART NOUVEAU STYLE, E. A. Seguy. 166 motifs, on 40 plates, from *Les fleurs et leurs applications decoratives* (1902): borders, circular designs, repeats, allovers, "spots." All in authentic Art Nouveau colors. 48pp. 9⅜ x 12¼. 23439-8 Pa. $5.00

A DIDEROT PICTORIAL ENCYCLOPEDIA OF TRADES AND INDUSTRY, edited by Charles C. Gillispie. 485 most interesting plates from the great French Encyclopedia of the 18th century show hundreds of working figures, artifacts, process, land and cityscapes; glassmaking, papermaking, metal extraction, construction, weaving, making furniture, clothing, wigs, dozens of other activities. Plates fully explained. 920pp. 9 x 12. 22284-5, 22285-3 Clothbd., Two-vol. set $40.00

HANDBOOK OF EARLY ADVERTISING ART, Clarence P. Hornung. Largest collection of copyright-free early and antique advertising art ever compiled. Over 6,000 illustrations, from Franklin's time to the 1890's for special effects, novelty. Valuable source, almost inexhaustible.
Pictorial Volume. Agriculture, the zodiac, animals, autos, birds, Christmas, fire engines, flowers, trees, musical instruments, ships, games and sports, much more. Arranged by subject matter and use. 237 plates. 288pp. 9 x 12. 20122-8 Clothbd. $14.50

Typographical Volume. Roman and Gothic faces ranging from 10 point to 300 point, "Barnum," German and Old English faces, script, logotypes, scrolls and flourishes, 1115 ornamental initials, 67 complete alphabets, more. 310 plates. 320pp. 9 x 12. 20123-6 Clothbd. $15.00

CALLIGRAPHY (CALLIGRAPHIA LATINA), J. G. Schwandner. High point of 18th-century ornamental calligraphy. Very ornate initials, scrolls, borders, cherubs, birds, lettered examples. 172pp. 9 x 13. 20475-8 Pa. $7.00

A MAYA GRAMMAR, Alfred M. Tozzer. Practical, useful English-language grammar by the Harvard anthropologist who was one of the three greatest American scholars in the area of Maya culture. Phonetics, grammatical processes, syntax, more. 301pp. 5⅜ x 8½. 23465-7 Pa. $4.00

THE JOURNAL OF HENRY D. THOREAU, edited by Bradford Torrey, F. H. Allen. Complete reprinting of 14 volumes, 1837-61, over two million words; the sourcebooks for *Walden*, etc. Definitive. All original sketches, plus 75 photographs. Introduction by Walter Harding. Total of 1804pp. 8½ x 12¼. 20312-3, 20313-1 Clothbd., Two-vol. set $70.00

CLASSIC GHOST STORIES, Charles Dickens and others. 18 wonderful stories you've wanted to reread: "The Monkey's Paw," "The House and the Brain," "The Upper Berth," "The Signalman," "Dracula's Guest," "The Tapestried Chamber," etc. Dickens, Scott, Mary Shelley, Stoker, etc. 330pp. 5⅜ x 8½. 20735-8 Pa. $4.50

SEVEN SCIENCE FICTION NOVELS, H. G. Wells. Full novels. *First Men in the Moon, Island of Dr. Moreau, War of the Worlds, Food of the Gods, Invisible Man, Time Machine, In the Days of the Comet.* A basic science-fiction library. 1015pp. 5⅜ x 8½. (Available in U.S. only)
20264-X Clothbd. $8.95

ARMADALE, Wilkie Collins. Third great mystery novel by the author of *The Woman in White* and *The Moonstone*. Ingeniously plotted narrative shows an exceptional command of character, incident and mood. Original magazine version with 40 illustrations. 597pp. 5⅜ x 8½.
23429-0 Pa. $6.00

MASTERS OF MYSTERY, H. Douglas Thomson. The first book in English (1931) devoted to history and aesthetics of detective story. Poe, Doyle, LeFanu, Dickens, many others, up to 1930. New introduction and notes by E. F. Bleiler. 288pp. 5⅜ x 8½. (Available in U.S. only)
23606-4 Pa. $4.00

FLATLAND, E. A. Abbott. Science-fiction classic explores life of 2-D being in 3-D world. Read also as introduction to thought about hyperspace. Introduction by Banesh Hoffmann. 16 illustrations. 103pp. 5⅜ x 8½.
20001-9 Pa. $2.00

THREE SUPERNATURAL NOVELS OF THE VICTORIAN PERIOD, edited, with an introduction, by E. F. Bleiler. Reprinted complete and unabridged, three great classics of the supernatural: *The Haunted Hotel* by Wilkie Collins, *The Haunted House at Latchford* by Mrs. J. H. Riddell, and *The Lost Stradivarious* by J. Meade Falkner. 325pp. 5⅜ x 8½.
22571-2 Pa. $4.00

AYESHA: THE RETURN OF "SHE," H. Rider Haggard. Virtuoso sequel featuring the great mythic creation, Ayesha, in an adventure that is fully as good as the first book, *She*. Original magazine version, with 47 original illustrations by Maurice Greiffenhagen. 189pp. 6½ x 9¼.
23649-8 Pa. $3.50

UNCLE SILAS, J. Sheridan LeFanu. Victorian Gothic mystery novel, considered by many best of period, even better than Collins or Dickens. Wonderful psychological terror. Introduction by Frederick Shroyer. 436pp. 5⅜ x 8½. 21715-9 Pa. $6.00

JURGEN, James Branch Cabell. The great erotic fantasy of the 1920's that delighted thousands, shocked thousands more. Full final text, Lane edition with 13 plates by Frank Pape. 346pp. 5⅜ x 8½. 23507-6 Pa. $4.50

THE CLAVERINGS, Anthony Trollope. Major novel, chronicling aspects of British Victorian society, personalities. Reprint of Cornhill serialization, 16 plates by M. Edwards; first reprint of full text. Introduction by Norman Donaldson. 412pp. 5⅜ x 8½. 23464-9 Pa. $5.00

KEPT IN THE DARK, Anthony Trollope. Unusual short novel about Victorian morality and abnormal psychology by the great English author. Probably the first American publication. Frontispiece by Sir John Millais. 92pp. 6½ x 9¼. 23609-9 Pa. $2.50

RALPH THE HEIR, Anthony Trollope. Forgotten tale of illegitimacy, inheritance. Master novel of Trollope's later years. Victorian country estates, clubs, Parliament, fox hunting, world of fully realized characters. Reprint of 1871 edition. 12 illustrations by F. A. Faser. 434pp. of text. 5⅜ x 8½. 23642-0 Pa. $5.00

YEKL and THE IMPORTED BRIDEGROOM AND OTHER STORIES OF THE NEW YORK GHETTO, Abraham Cahan. Film *Hester Street* based on *Yekl* (1896). Novel, other stories among first about Jewish immigrants of N.Y.'s East Side. Highly praised by W. D. Howells—Cahan "a new star of realism." New introduction by Bernard G. Richards. 240pp. 5⅜ x 8½. 22427-9 Pa. $3.50

THE HIGH PLACE, James Branch Cabell. Great fantasy writer's enchanting comedy of disenchantment set in 18th-century France. Considered by some critics to be even better than his famous *Jurgen*. 10 illustrations and numerous vignettes by noted fantasy artist Frank C. Pape. 320pp. 5⅜ x 8½. 23670-6 Pa. $4.00

ALICE'S ADVENTURES UNDER GROUND, Lewis Carroll. Facsimile of ms. Carroll gave Alice Liddell in 1864. Different in many ways from final Alice. Handlettered, illustrated by Carroll. Introduction by Martin Gardner. 128pp. 5⅜ x 8½. 21482-6 Pa. $2.50

FAVORITE ANDREW LANG FAIRY TALE BOOKS IN MANY COLORS, Andrew Lang. The four Lang favorites in a boxed set—the complete *Red, Green, Yellow* and *Blue* Fairy Books. 164 stories; 439 illustrations by Lancelot Speed, Henry Ford and G. P. Jacomb Hood. Total of about 1500pp. 5⅜ x 8½. 23407-X Boxed set, Pa. $15.95

AN AUTOBIOGRAPHY, Margaret Sanger. Exciting personal account of hard-fought battle for woman's right to birth control, against prejudice, church, law. Foremost feminist document. 504pp. 5⅜ x 8½.
20470-7 Pa. $5.50

MY BONDAGE AND MY FREEDOM, Frederick Douglass. Born as a slave, Douglass became outspoken force in antislavery movement. The best of Douglass's autobiographies. Graphic description of slave life. Introduction by P. Foner. 464pp. 5⅜ x 8½. 22457-0 Pa. $5.50

LIVING MY LIFE, Emma Goldman. Candid, no holds barred account by foremost American anarchist: her own life, anarchist movement, famous contemporaries, ideas and their impact. Struggles and confrontations in America, plus deportation to U.S.S.R. Shocking inside account of persecution of anarchists under Lenin. 13 plates. Total of 944pp. 5⅜ x 8½.
22543-7, 22544-5 Pa., Two-vol. set $12.00

LETTERS AND NOTES ON THE MANNERS, CUSTOMS AND CONDITIONS OF THE NORTH AMERICAN INDIANS, George Catlin. Classic account of life among Plains Indians: ceremonies, hunt, warfare, etc. Dover edition reproduces for first time all original paintings. 312 plates. 572pp. of text. 6⅛ x 9¼. 22118-0, 22119-9 Pa.. Two-vol. set $12.00

THE MAYA AND THEIR NEIGHBORS, edited by Clarence L. Hay, others. Synoptic view of Maya civilization in broadest sense, together with Northern, Southern neighbors. Integrates much background, valuable detail not elsewhere. Prepared by greatest scholars: Kroeber, Morley, Thompson, Spinden, Vaillant, many others. Sometimes called Tozzer Memorial Volume. 60 illustrations, linguistic map. 634pp. 5⅜ x 8½.
23510-6 Pa. $10.00

HANDBOOK OF THE INDIANS OF CALIFORNIA, A. L. Kroeber. Foremost American anthropologist offers complete ethnographic study of each group. Monumental classic. 459 illustrations, maps. 995pp. 5⅜ x 8½.
23368-5 Pa. $13.00

SHAKTI AND SHAKTA, Arthur Avalon. First book to give clear, cohesive analysis of Shakta doctrine, Shakta ritual and Kundalini Shakti (yoga). Important work by one of world's foremost students of Shaktic and Tantric thought. 732pp. 5⅜ x 8½. (Available in U.S. only)
23645-5 Pa. $7.95

AN INTRODUCTION TO THE STUDY OF THE MAYA HIEROGLYPHS, Syvanus Griswold Morley. Classic study by one of the truly great figures in hieroglyph research. Still the best introduction for the student for reading Maya hieroglyphs. New introduction by J. Eric S. Thompson. 117 illustrations. 284pp. 5⅜ x 8½. 23108-9 Pa. $4.00

A STUDY OF MAYA ART, Herbert J. Spinden. Landmark classic interprets Maya symbolism, estimates styles, covers ceramics, architecture, murals, stone carvings as artforms. Still a basic book in area. New introduction by J. Eric Thompson. Over 750 illustrations. 341pp. 8⅜ x 11¼.
21235-1 Pa. $6.95